Introduction to Microprocessors and Microcontrollers

Introduction to Microprocessors and Microcontrollers

Second edition

John Crisp

AMSTERDAM • BOSTON • HEIDELBERG • LONDON • NEW YORK • OXFORD
PARIS • SAN DIEGO • SAN FRANCISCO • SINGAPORE • SYDNEY • TOKYO

Newnes is an imprint of Elsevier

Newnes
An imprint of Elsevier
Linacre House, Jordan Hill, Oxford OX2 8DP
200 Wheeler Road, Burlington MA 01803

First published 1998 as *Introduction to Microprocessors*
Reprinted 2000, 2001
Second Edition 2004

British Library Cataloguing in Publication Data
A catalogue record for this book is available from the British Library

ISBN 0 7506 5989 0

For information on all Newnes publications
visit our website at: www.newnespress.com

Composition by Genesis Typesetting Limited, Rochester, Kent
Printed and bound in Great Britain

Contents

Preface		vii
1	Basic microprocessor systems	1
2	Binary – the way micros count	8
3	Hexadecimal – the way we communicate with micros	25
4	How micros calculate	38
5	An introduction to logic gates and their uses	49
6	Registers and memories	62
7	A microprocessor-based system	85
8	A typical 8-bit microprocessor	99
9	Programming – using machine code and assembly language	121
10	High-level languages	132
11	The development of microprocessors and microcontrollers	151
12	The Pentium family	173
13	The PowerPC	184
14	The Athlon XP	194
15	Microcontrollers and how to use them	199
16	Using a PIC microcontroller for a real project	219
17	Interfacing	234
18	Test equipment and fault-finding	255
Appendix A: Special function register file		267
Appendix B: PIC 16CXXX instruction set		268
Further reading		271
Quiz time answers		273
Index		275

Preface

The first edition of this book started with the words: 'A modern society could no longer function without the microprocessor.'

This is certainly still true but it is even truer if we include the microcontroller. While the microprocessor is at the heart of our computers, with a great deal of publicity, the microcontroller is quietly running the rest of our world. They share our homes, our vehicles and our workplace, and sing to us from our greetings cards.

They are our constant, unseen companions and billions are being installed every year with little or no publicity.

The purpose of this book is to give a worry-free introduction to microprocessors and microcontrollers. It starts at the beginning and does not assume any previous knowledge of microprocessors or microcontrollers and, in gentle steps, introduces the knowledge necessary to take those vital first steps into the world of the micro.

John Crisp

1

Basic microprocessor systems

The microprocessor was born

In 1971 two companies, both in the USA, introduced the world to its future by producing microprocessors. They were a young company called Intel and their rival, Texas Instruments.

The microprocessor and its offspring, the microcontroller, were destined to infiltrate every country, every means of production, and almost every home in the world. There is now hardly a person on the planet that does not own or know of something that is dependent on one of these devices. Yet curiously, so few people can give any sort of answer to the simple question 'What is a microprocessor?' This, and 'How does it work?' form two of the starting points for this book.

Let's start by looking at a system

The word 'system' is used to describe any organization or device that includes three features.

A system must have at least one input, one output and must do something, i.e. it must contain a process. Often there are many inputs and outputs. Some of the outputs are required and some are waste products. To a greater or lesser extent, all processes generate some waste heat. Figure 1.1 shows these requirements.

Figure 1.1

The essential requirements of a system

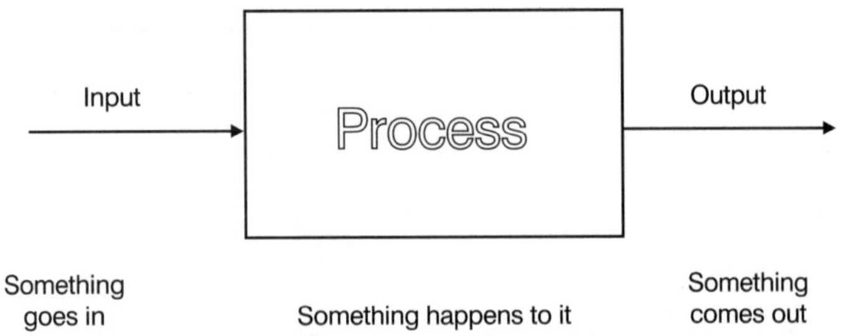

Input → Process → Output

Something goes in   Something happens to it   Something comes out

A wide range of different devices meets these simple requirements. For example, a motor car will usually require fuel, water for cooling purposes and a battery to start the engine and provide for the lights and instruments. Its process it to burn the fuel and extract the energy to provide transportation for people and goods. The outputs are the wanted movement and the unwanted pollutants such as gases, heat, water vapour and noise.

The motor car contains other systems within it. In Figure 1.2, we added electricity as a required input to start the engine and provide the lights

Figure 1.2

An everyday system

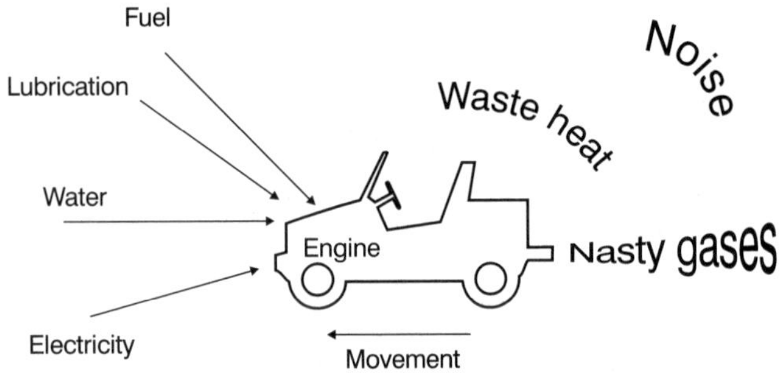

Fuel
Lubrication
Water
Electricity
Engine
Waste heat
Noise
Nasty gases
Movement

Figure 1.3

Recharging the battery

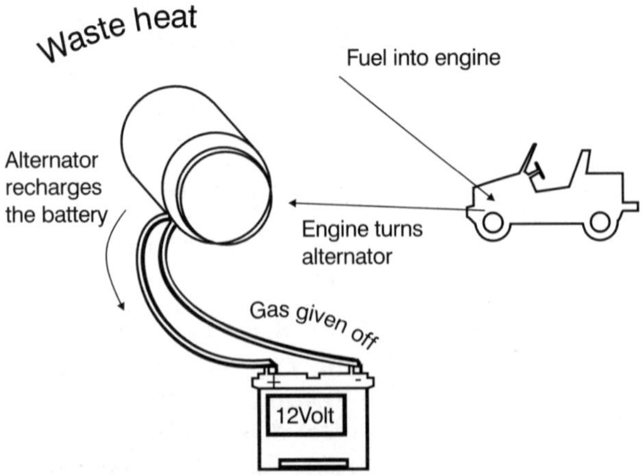

Waste heat
Fuel into engine
Alternator recharges the battery
Engine turns alternator
Gas given off
12Volt

and the instruments but thereafter the battery is recharged by the engine. There must, then, be an electrical system at work, as in Figure 1.3, so it is quite possible for systems to have smaller systems inside or embedded within them. In a similar way, a motor car is just a part of the transport system.

A microprocessor system

Like any other system, a microprocessor has inputs, outputs and a process as shown in Figure 1.4. The inputs and outputs of a microprocessor are a series of voltages that can be used to control external devices. The process involves analysing the input voltages and using them to 'decide' on the required output voltages. The decision is based on previously entered instructions that are followed quite blindly, sensible or not.

Figure 1.4

The microprocessor system

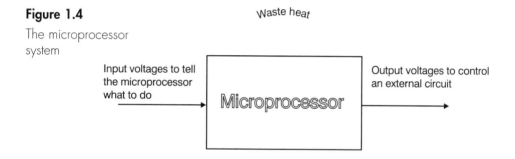

Waste heat

Input voltages to tell the microprocessor what to do

Microprocessor

Output voltages to control an external circuit

His and hers garage door opener

Here is a little task that a simple microprocessor can solve for us. When the woman arrives in her car, a light signal is flashed at the sensor and only her garage door opens. When the man arrives home, his car flashes a light signal at the same sensor but this time his garage door opens but hers remains closed.

The cars are sending a different sequence of light flashes to the light sensor. The light sensor converts the incoming light to electrical voltage pulses that are recognized by the microprocessor. The output voltage now operates the electrical motor attached to the appropriate door. The overall scheme is shown in Figure 1.5.

In the unlikely event of it being needed, a modern microprocessor would find it an easy task to increase the number of cars and garages to include every car and every garage that has ever been manufactured. Connecting all the wires, however, would be an altogether different problem!

3

Figure 1.5

Opening the right garage door

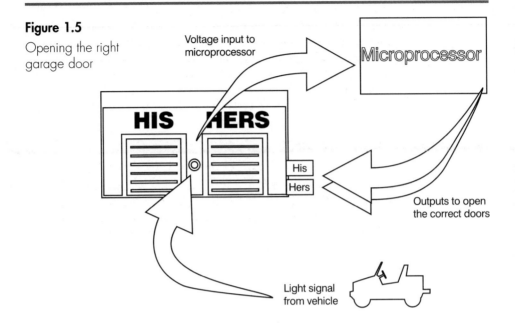

The physical appearance of a microprocessor

A microprocessor is a very small electronic circuit typically ½ inch (12 mm) across. It is easily damaged by moisture or abrasion so to offer it some protection it is encapsulated in plastic or ceramic. To provide electrical connections directly to the circuit would be impractical owing to the size and consequent fragility, so connecting pins are moulded into the case and the microprocessor then plugs into a socket

Figure 1.6

Typical microprocessors

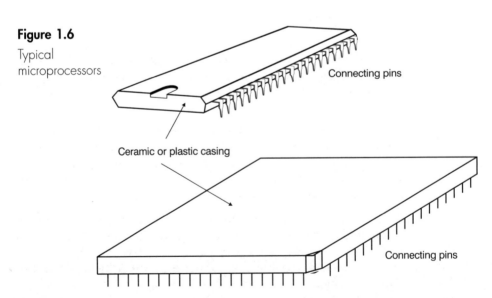

on the main circuit board. The size, shape and number of pins on the microprocessor depend on the amount of data that it is designed to handle. The trend, as in many fields, is forever upward. Typical microprocessors are shown in Figure 1.6.

Terminology

Integrated circuits

An electronic circuit fabricated out of a solid block of semiconductor material. This design of circuit, often called a solid state circuit, allows for very complex circuits to be constructed in a small volume. An integrated circuit is also called a 'chip'.

Microprocessor (μp)

This is the device that you buy: just an integrated circuit as in Figure 1.6. On its own, without a surrounding circuit and applied voltages it is quite useless. It will just lie on your workbench staring back at you.

Microprocessor-based system

This is any system that contains a microprocessor, and does not necessarily have anything to do with computing. In fact, despite all the hype, computers use only a small proportion of all the micro-processors manufactured. Our garage door opening system is a microprocessor-based system or is sometimes called a microprocessor-controlled system.

Microcomputer

The particular microprocessor-based systems that happen to be used as a computer are called microcomputers. The additional circuits required for a computer can be built into the same integrated circuit giving rise to a single chip microcomputer.

Microcontroller

This is a complete microprocessor-based control system built onto a single chip. It is small and convenient but doesn't do anything that could not be done with a microprocessor and a few additional components. We'll have a detailed look at these in a later chapter.

MPU and CPU

An MPU is a MicroProcessor Unit or microprocessor. A CPU is a Central Processing Unit. This is the central 'brain' of a computer and

5

can be (usually is) made from one or more microprocessors. The IBM design for the 'Blue Gene' supercomputer includes a million processors!

Remember:

MPU is the thing
CPU is the job.

Micro

The word micro is used in electronics and in science generally, to mean 'one-millionth' or 1×10^{-6}. It has also entered general language to mean something very small like a very small processor or microprocessor. It has also become an abbreviation for microprocessor, microcomputer, microprocessor-based system or a micro controller – indeed almost anything that has 'micro' in its name. In the scientific sense, the word micro is represented by the Greek letter μ (mu). It was only a small step for microprocessor to become abbreviated to μP.

Some confusion can arise unless we make sure that everyone concerned is referring to the same thing.

Quiz time 1

In each case, choose the best option.

1 A microprocessor:

(a) requires fuel, water and electricity.
(b) is abbreviated to μc.
(c) is often encapsulated in plastic.
(d) is never used in a CPU but can be used in an MPU.

2 A system must include:

(a) an input, an output and a process.
(b) something to do with a form of transport.
(c) a microprocessor.
(d) fuel, water and electricity.

3 All systems generate:

(a) movement.
(b) chips.
(c) waste heat.
(d) waste gases.

4 An MPU:

(a) is the same as a μP.
(b) can be made from more than one Central Processing Unit.
(c) is a small, single chip computer.
(d) is an abbreviation for Main Processing Unit.

5 Integrated circuits are *not*:

(a) called chips.
(b) used to construct a microprocessor-based system.
(c) solid state circuits.
(d) an essential part of an engine.

2

Binary – the way micros count

Unlike us, microprocessors have not grown up with the idea that 10 is a convenient number of digits to use. We have taken it so much for granted that we have even used the word digit to mean both a finger and a number.

Microprocessors and other digital circuits use only two digits – 0 and 1 – but why? Ideally, we would like our microprocessors to do everything at infinite speed and never make a mistake. Error free or high speed – which would you feel is the more important?

It's your choice but I would go for error free every time, particularly when driving my car with its engine management computer or when coming in to land in a fly-by-wire aircraft. I think most people would agree.

So let's start by having a look at one effect of persuading microprocessors to count in our way.

The noise problem

If the input of a microprocessor is held at a constant voltage, say 4 V, this would appear as in Figure 2.1.

If we try to do this in practice, then careful measurements would show that the voltage is not of constant value but is continuously wandering

Figure 2.1

A constant voltage

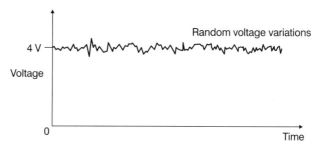

above and below the mean level. These random fluctuations are called electrical noise and degrade the performance of every electronic circuit. We can take steps to reduce the effects but preventing it altogether is, so far, totally impossible. We can see the effect by disconnecting the antenna of our television. The noise causes random speckles on the screen which we call snow. The same effect causes an audible hiss from the loudspeaker. The effect of noise is shown in Figure 2.2.

Figure 2.2

A 'noisy' voltage

Most microprocessors use a power supply of 5 V or 3.3 V. To keep the arithmetic easy, we will assume a 5 V system.

If we are going to persuade the microprocessor to count from 0 to 9, as we do, using voltages available on a 5 V supply would give 0.5 V per digit:

$$0 = 0\,\text{V}$$
$$1 = 0.5\,\text{V}$$
$$2 = 1\,\text{V}$$
$$3 = 1.5\,\text{V}$$
$$4 = 2\,\text{V}$$
$$5 = 2.5\,\text{V}$$
$$6 = 3\,\text{V}$$
$$7 = 3.5\,\text{V}$$
$$8 = 4\,\text{V}$$
$$9 = 4.5\,\text{V}$$

If we were to instruct our microprocessor to perform the task 4 + 4 = 8, by pressing the '4' key we could generate a 2 V signal which is then remembered by the microprocessor. The + key would tell it to add and pressing the '4' key again would then generate another 2 V signal.

So, inside the microprocessor we would see it add the 2 V and then another 2 V and, hence, get a total of 4 V. The microprocessor could then use the list shown to convert the total voltage to the required numerical result of 8. This simple addition is shown in Figure 2.3.

This seemed to work nicely – but we ignored the effect of noise. Figure 2.4 shows what could happen. The exact voltage memorized by the microprocessor would be a matter of chance. The first time we pressed

Figure 2.3

It works! 4 + 4 does equal 8

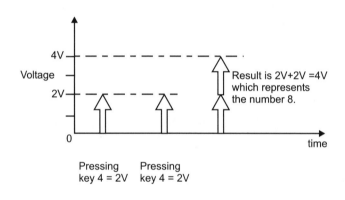

Figure 2.4

Noise can cause problems

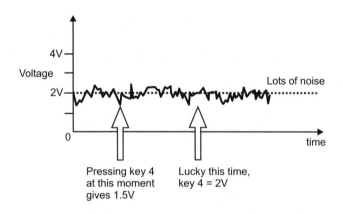

key 4, the voltage just happened to be at 1.5 V but the second time we were luckier and the voltage was at the correct value of 2 V.

Inside the microprocessor:

$$1.5 V + 2 V = 3.5 V$$

and using the table, the 3.5 V is then converted to the number 7. So our microprocessor reckons that 4 + 4 = 7.5!

Since the noise is random, it is possible, of course, to get a final result that is too low, too high or even correct.

A complete cure for electrical noise

Sorry, just dreaming. There isn't one. The small particle-like components of electricity, called electrons, vibrate in a random fashion powered by the surrounding heat energy. In conductors, electrons are very mobile and carry a type of electrical charge that we have termed negative. The resulting negative charge is balanced out by an equal number of fixed particles called protons, which carry a positive charge (see Figure 2.5).

The overall effect of the electron mobility is similar to the random surges that occur in a large crowd of people jostling around waiting to enter the stadium for the Big Match. If, at a particular time, there happens to be more electrons or negative charges moving towards the

Figure 2.5

Equal charges result in no overall voltage

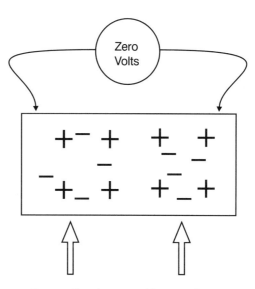

Four positive charges and four negative charges at each end – therefore no voltage difference between the ends

Figure 2.6

A random voltage has been generated

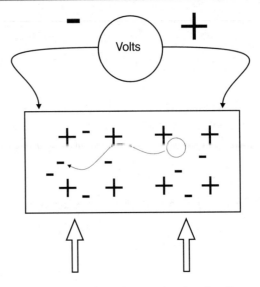

One negative charge has wandered up the other end making the left-hand end more negative

left-hand end of a piece of material then that end would become more negative, as shown in Figure 2.6. A moment later, the opposite result may occur and the end would become more positive (Figure 2.7). These effects give rise to small random voltages in any conductor, as we have seen.

Figure 2.7

The opposite effect is equally likely

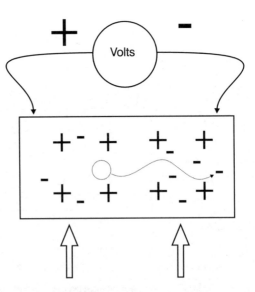

If it happened to move the other way, the right-hand end would be negative

Thermal noise

The higher the temperature, the more mobile the electrons, the greater the random voltages and the more electrical noise is present.

A solution:

High temperature = high noise

so:

Low temperature = low noise.

Put the whole system into a very cold environment by dropping it in liquid nitrogen (about −200°C) or taking it into space where the 'shade' temperature is about −269°C. The cold of space has created very pleasant low noise conditions for the circuits in space like the Hubble telescope. On Earth most microprocessors operate at room temperature. It would be inconvenient, not to mention expensive, to surround all our microprocessor circuits by liquid nitrogen. And even if we did, there is another problem queuing up to take its place.

Partition noise

Let's return to the Big Match. Two doors finally open and the fans pour through the turnstiles. Now we may expect an equal number of people to pass through the two entrances as shown in Figure 2.8 but in reality this will not happen. Someone will have trouble finding their ticket; friends will wait for each other; cash will be offered instead of a ticket; someone will try to get back out through the gate to reach another section of the stadium. As we can imagine, the streams of people may be equal over an hour but second by second random fluctuations will occur.

Electrons don't lose their tickets but random effects like temperature, voltage and interactions between adjacent electrons have a very similar effect.

Figure 2.8

The fans enter the stadium

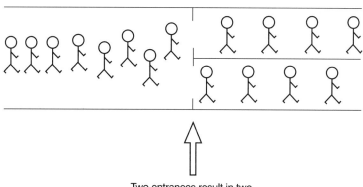

Two entrances result in two
equal streams of people

A single current of, say, 1 A can be split into two currents of 0.5 A when measured over the long-term, but when examined carefully, each will contain random fluctuations. This type of electrical noise is called partition noise or partition effect. The overall effect is similar to the thermal noise and, between them, would cause too much noise and hence would rule out the use of a 10-digit system.

How much noise can we put up with?

The 10-finger system that we use is called a 'denary' or 'decimal' system. We have seen that a 5 V supply would accommodate a 10-digit counting system if each digit was separated by 0.5 V or, using the more modern choice of 3.3 V, the digits would be separated by only 0.33 V.

> *Question*: Using a 5 V supply and a denary system, what is the highest noise voltage that can be tolerated?

> *Answer*: Each digit is separated by only 5 V/10 = 0.5 V.

The number 6 for example would have a value of 3 V and the number 7 would be represented by 3.5 V. If the noise voltage were to increase the 3 V to over 3.25 V, the number is likely to be misread as 7. The highest acceptable noise level would therefore be 0.25 V. This is not very high and errors would be common. If we used a supply voltage of 3.3 V, the situation would get even worse.

So why don't we just increase the operating voltage to say, 10 V, or 100 V? The higher the supply voltage the less likely it is that electrical noise would be a problem. This is true but the effect of increasing the supply would be to require thicker insulation and would increase the physical size of the microprocessor and reduce its speed. More about this in Chapter 11.

Using just two digits

If we reduce the number of digits then a wider voltage range can be used for each value and the errors due to noise are likely to occur less often.

We have chosen to use only two digits, 0 and 1, to provide the maximum degree of reliability. A further improvement is to provide a safety zone between each voltage. Instead of taking our supply voltage of 3.3 V and simply using the lower half to represent the digit 0 and the top half for 1, we allocate only the lower third to 0 and the upper third to 1 as shown in Figure 2.9. This means that the noise level will have to be at least 1.1 V (one-third of 3.3 V) to push a level 0 digit up to the minimum value for a level 1.

Figure 2.9

A better choice of voltages

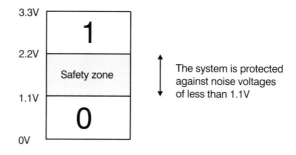

The voltages chosen to represent the digits 0 and 1

How do we count?

Normally, we count in the system we call 'denary'. We start with 0 then go to 1 then to a new symbol that we write as 2 and call 'two'. This continues until we run out of symbols. So far, it looks like this:

0
1
2
3
4
5
6
7
8
9

At this point we have used all the symbols once and, to show this, we put a '1' to the left of the numbers as we re-use them. This gives us:

10
11
12
13
14

. . . and so on up to 19 when we put a 2 on the left-hand side and start again 20, 21, 22 etc.

When we reach 99, we again add a '1' on the left-hand side and put the other digits back to zero to give 100. After we reach 999, we go to 1000 and so on.

Counting is not easy. We often take it for granted but if we think back to our early days at school, it took the teacher over a year before we were happy and reasonably competent. So counting is more difficult than microprocessors – you've mastered the difficult part already!

15

The basic basis of bases

The base of a number system is the number of different symbols used in it. In the case of the denary system, we use 10 different symbols, 0 . . . 9, other numbers, like 28 657, are simply combinations of the 10 basic symbols.

Since the denary system uses 10 digits, the system is said to have a base of 10. The base is therefore just the technical word for the number of digits used in any counting system.

Counting with only two figures

We can count using any base that we like. In the denary or decimal system, we used a base of 10 but we have seen that microprocessors use a base of 2 – just the two digits 0 and 1. This is called the binary system.

We usually abbreviate the words BInary digiT to bit.

Counting follows the same pattern as we have seen in the denary system: we use up the digits then start again.

Let's give it a try. Start by listing all the digits:

0
1

and that's it!

We now put a '1' in the next column and start again:

10
11

It is convenient at this stage to keep the number of binary columns the same and so we add a 0 at the start of the first two digits. These extra zeros do not alter the value at all. For example, the denary number 25 is not affected by writing it as 025 or 0025 or even 000 000 000 000 025.

The binary and decimal equivalents are:

Binary	Denary
00	0
01	1
10	2
11	3

We do the same again – put a '1' in the next column and repeat the pattern to give:

Binary	Denary
100	4
101	5
110	6
111	7

and once more:

Binary	Denary
1000	8
1001	9
1010	10
1011	11

Confusion and the cure

Here is a number: 1000. But what number is it? Is it a thousand in denary or is it eight written in binary?

I don't know. I could take a guess but the difference between flying an aircraft at eight feet and a thousand feet is a serious matter. The only way to be certain is to say so at the time. This is done by showing the base of the number system being used to make the meaning quite clear. The base of the number system is shown as a subscript after the number.

If the 1000 were a binary number, it is written as 1000_2 and if it were a denary number it would be shown as 1000_{10}.

It would be easy to advise that the base of the number system in use is always shown against every number but this would be totally unrealistic. No one is going to write a base after their telephone number or a price in a shop. Use a base when it would be useful to avoid confusion, such as by writing statements like 1000 = 8 (a thousand = eight???). Write it as $1000_2 = 8_{10}$ and make life a little easier.

Converting denary to binary

Of course, if someone were to ask us for the binary equivalent of nine we could just start from zero and count up until we reach nine. This is a boring way to do it and with larger numbers like $1\,000\,000_{10}$ it would be very tedious indeed. Here is a better way. The method will be explained using the conversion of 52_{10} to binary as an example.

17

A worked example

Convert 52_{10} to binary

Step 1: Write down the number to be converted

      52

Step 2: Divide it by 2 (because 2 is the base of the binary system), write the whole number part of the answer underneath and the remainder 0 or 1 alongside

      52
      26 0

Step 3: Divide the answer (26) by 2 and record the remainder (0) as before

      52
      26 0
      13 0

Step 4: Divide the 13 by 2 and write down the answer (6) and the remainder (1)

      52
      26 0
      13 0
       6 1

Step 5: 2 into 6 goes 3 remainder 0

      52
      26 0
      13 0
       6 1
       3 0

Step 6: Dividing 3 gives an answer of 1 and a remainder of 1

      52
      26 0
      13 0
       6 1
       3 0
       1 1

Step 7: Finally, dividing the 1 by 2 will give 0 and a remainder of 1

      52
      26 0
      13 0
       6 1
       3 0
       1 1
       0 1

Step 8: We cannot go any further with the divisions because all the answers will be zero from now on. The binary number now appears in the remainder column. To get the answer read the remainder column from the bottom UPWARDS

```
52
26   0        = 110100₂
13   0   ↑
 6   1
 3   0
 1   1
 0   1
```

Method

1 Divide the denary number by 2 – write the whole number result underneath and the remainder in a column to the right.

2 Repeat the process until the number is reduced to zero.

3 The binary number is found by reading the remainder column from the bottom upwards.

Another example

Here is one for you to try. If you get stuck, the solution is given below.

Convert 2187_{10} to a binary number

```
2187
1093   1   = 100010001011₂
 546   1           ↑
 273   0
 136   1
  68   0
  34   0
  17   0
   8   1
   4   0
   2   0
   1   0
   0   1
```

Doing it by calculator: Many scientific calculators can do the conversion of denary to binary for us. Unfortunately, they are limited to quite low numbers by the number of digits able to be seen on the screen.

To do a conversion, we need:

1 A scientific calculator that can handle different number bases.
2 The instruction booklet.

3 About half an hour to spare – or a week if you have lost the instructions.

The exact method varies but on my elderly Casio it goes something like this:

To tell the calculator that the answer has to be in binary I have to press mode mode 3 then the 'binary' key.

It now has to be told that the input number is decimal. This is the exciting key sequence logic logic logic 1 now just put in our number 52 and press the = key and out will pop the answer 110100.

Converting binary to denary

If we look at a denary number like 8328, we see that it contains two eights. Now these two figures look identical however closely we examine them, but we know that they are different. The 8 on the right-hand end is really 8 but the other one is actually 8000 because it is in the thousands column.

The real value of a digit is dependent on two things: the digit used and the column in which it is placed.

In the denary system, the columns, starting from the right, are units, tens, hundreds, thousands etc. Rather than use these words, we could express them in powers of ten. A thousand is $10 \times 10 \times 10 = 10^3$ and in a similar way, a hundred is 10^2, ten is 10^1 and a unit is 10^0. Each column simply increases the power applied to the base of the number system.

Columns in a binary world also use the base raised to increasing powers as we move across the columns towards the left.

So we have:

$$2^3 \quad 2^2 \quad 2^1 \quad 2^0$$

The denary equivalent can be found by multiplying out the powers of two. So 2^3 is $2 \times 2 \times 2 = 8$ and $2^2 = 4$, $2^1 = 2$ and finally $2^0 = 1$. Starting from the right-hand side, the column values would be 1, 2, 4, 8 etc. Let's use this to convert the binary number 1001 into denary.

Method

Step 1: Write down the values of the columns

8 4 2 1

Step 2: Write the binary number underneath

8 4 2 1
1 0 0 1

Step 3: Evaluate the values of the columns

$$8 \times 1 = 8$$
$$4 \times 0 = 0$$
$$2 \times 0 = 0$$
$$1 \times 1 = 1$$

Step 4: Add up the values

$$8 + 1 = 9$$

As we have seen, all the columns containing a binary 0 can be ignored because they always come out to 0 so a quicker way is to simply add up all the column values where the binary digit is 1.

Method

1 Write down the column values for the binary system using the same number of columns as are shown in the binary number.

2 Enter the binary number, one bit under each column heading.

3 Add the values of each column where a 1 appears in the binary number.

Calculator note: This is much the same as we saw the previous conversion. To tell the calculator that the answer has to be in decimal I have to press mode mode 3 then the 'decimal' key.

It now has to be told that the input number is binary. This is done by the key sequence logic logic logic 3 now just put in our binary number 1001 and press the = key and out will pop the answer 9.

Another example

Once again, here is one for you to try. If you have problems, the answer follows.

Convert 101100101_2 to a denary number

Step 1: Write down the column values by starting with a 1 on the right-hand side then just keep doubling as necessary

2^8	2^7	2^6	2^5	2^4	2^3	2^2	2^1	2^0
256	128	64	32	16	8	4	2	1

Step 2: Enter the binary number under the column headings

256	128	64	32	16	8	4	2	1
1	0	1	1	0	0	1	0	1

Step 3: Add up all the column values where the binary digit is 1

$$256 + 64 + 32 + 4 + 1 = 357$$

So, $101100101_2 = 357_{10}$ or just 357 since denary can be assumed in this case.

Bits, bytes and other things

All the information entering or leaving a microprocessor is in the form of a binary signal, a voltage switching between the two bit levels 0 and 1.

Bits are passed through the microprocessor at very high speed and in large numbers and we find it easier to group them together.

Nibble

A group of four bits handled as a single lump. It is half a byte.

Byte

A byte is simply a collection of 8 bits. Whether they are ones or zeros or what their purpose is does not matter.

Word

A number of bits can be collected together to form a 'word'. Unlike a byte, a word does not have a fixed number of bits in it. The length of the word or the number of bits in the word depends on the microprocessor being used.

If the microprocessor accepts binary data in groups of 32 at a time then the word in this context would include 32 bits. If a different microprocessor used data in smaller handfuls, say 16 at a time, then the word would have a value of 16 bits. The word is unusual in this context in as much as its size or length will vary according to the situations in which it is discussed. The most likely values are 8, 16, 32 and 64 bits but no value is excluded.

Long word

In some microprocessors where a word is taken to mean say 16 bits, a long word would mean a group of twice the normal length, in this case 32 bits.

Kilobyte (Kb or KB or kbyte)

A kilobyte is 1024 or 2^{10} bytes. In normal use, kilo means 1000 so a kilovolt or kV is exactly 1000 volts. In the binary system, the nearest column value to 1000 is 1024 since $2^9 = 512$ and $2^{10} = 1024$.

The difference between 1000 and 1024 is fairly slight when we have only 1 or 2 Kb and the difference is easily ignored. However, as the numbers increase, so does the difference. The actual number of bytes in 42 Kb is actually 43 008 bytes (42 $\times$ 1024).

The move in the computing world to use an upper case K to mean 1024 rather than k for meaning 1000 is trying to address this problem.

Unfortunately, even the upper or lower case b is not standardized so tread warily and look for clues to discover which value is being used. If in doubt use 1024 if it is to do with microprocessors or computers.

Bits often help to confuse the situation even further. 1000 bits is a kilobit or kb. Sometimes 1024 bits is a Kb. One way to solve the bit/byte problem is to use kbit (or Kbit) and kbyte (or Kbyte).

Megabyte (MB or Mb)

This is a kilokilobyte or 1024 $\times$ 1024 bytes. Numerically this is 2^{20} or 1 048 576 bytes. Be careful not to confuse this with mega as in megavolts (MV) which is exactly one million (10^6).

Gigabyte (Gb)

This is 1024 megabytes which is 2^{30} or 1 073 741 824 bytes. In general engineering, giga means one thousand million (10^9).

Terabyte (TB or Tb)

Terabyte is a megamegabyte or 2^{40} or 1 099 511 600 000 bytes (Tera = 10^{12}).

Petabyte (PB or Pb)

This is a thousand (or 1024) times larger than the Terabyte so it is 10^{15} in round numbers or 2^{40} which is pretty big. If you are really interested, you can multiply it out yourself by multiplying the TB figure by 1024.

Quiz time 2

In each case, choose the best option.

---

1   Typical operating voltages of microprocessors are:

(a) 0 V and 1 V.
(b) 3.3 V and 5 V.
(c) 220 V
(d) 1024 V.

---

2   The most mobile electrical charge is called:

(a) a proton and has a positive charge.
(b) a voltage and is always at one end of a conductor.
(c) an electron and has a negative charge.
(d) an electron and has a positive charge.

3 The denary number 600 is equivalent to the binary number:

(a) 1001011000.
(b) 011000000000.
(c) 1101001.
(d) 1010110000.

4 When converted to a denary number, the binary number 110101110:

(a) will end with a 0.
(b) must be greater than 256 but less than 512.
(c) will have a base of 2.
(d) will equal 656.

5 A byte:

(a) is either 1024 or 1000 bits.
(b) is simply a collection of 16 bits.
(c) can vary in length according to the microprocessor used.
(d) can have the same number of bits as a word.

3

Hexadecimal – the way we communicate with micros

The only problem with binary

The only problem with binary is that we find it so difficult and make too many errors. There is little point in designing microprocessors to handle binary numbers at high speed and with almost 100% accuracy if we are going to make loads of mistakes putting the numbers in and reading the answers.

From our point of view, binary has two drawbacks: the numbers are too long and secondly they are too tedious. If we have streams and streams of ones and zeros we get bored, we lose our place and do sections twice and miss bits out.

The speed of light in m/s can be written in denary as 299792459_{10} or in binary as $10001110111100111100001001011_2$. Try writing these numbers on a sheet of paper and we can be sure that the denary number will be found infinitely easier to handle. Incidentally, this binary number is less than half the length that a modern microprocessor can handle several millions of times a second with (almost) total accuracy.

In trying to make a denary number even easier, we tend to split it up into groups and would write or read it as 299 792 459. In this way, we are dealing with bite-sized portions and the 10 different digits ensure that there is enough variety to keep us interested.

We can perform a similar trick with binary and split the number into groups of four bits starting from the right-hand end as we do with denary numbers.

$$1 \quad 0001 \quad 1101 \quad 1110 \quad 0111 \quad 1000 \quad 0100 \quad 1011$$

Already it looks more digestible.

Now, if we take a group of four bits, the lowest possible value is 0000_2 and the highest is 1111_2. If these binary numbers are converted to denary, the possibilities range from 0 to 15.

Hexadecimal, or 'hex' to its friends

Counting from 0 to 15 will mean 16 different digits and so has a base of 16. What the digits look like really doesn't matter. Nevertheless, we may as well make it as simple as possible.

The first 10 are easy, we can just use 0123456789 as in denary. For the last six we have decided to use the first six letters of the alphabet: ABCDEF or abcdef.

The hex system starts as:

Hex	Denary
0	0
1	1
2	2
3	3
4	4
5	5
6	6
7	7
8	8
9	9
A	10
B	11
C	12
D	13
E	14
F	15

When we run out of digits, we just put a 1 in the second column and reset the first column to zero just as we always do.

So the count will continue:

10	16
11	17
12	18
13	19
14	20
15	21
16	22
17	23
18	24
19	25
1A	26
1B	27
1C	28
1D	29
1E	30
1F	31
20	32

. . . and so on.

It takes a moment or two to get used to the idea of having numbers that include letters but it soon passes. We must be careful to include the base whenever necessary to avoid confusion. The base is usually written as H, though h or 16 would still be acceptable.

'One eight' in hex is equal to twenty-four in denary. Notice how I avoided quoting the hex number as eighteen. Eighteen is a denary number and does not exist in hex. If you read it in this manner it reinforces the fact that it is not a denary value.

Here are the main options in order of popularity:

$$16H = 24_{10}$$
$$16_H = 24_{10}$$
$$16h = 24_{10}$$
$$16_h = 24_{10}$$
$$16_{16} = 24_{10}$$

The advantages of hex

1 It is very compact. Using a base of 16 means that the number of digits used to represent a given number is usually fewer than in binary or denary.
2 It is easy to convert between hex and binary and fairly easy to go between hex and denary. Remember that the microprocessor only works in binary, all the conversions between hex and binary are carried out in other circuits (Figure 3.1).

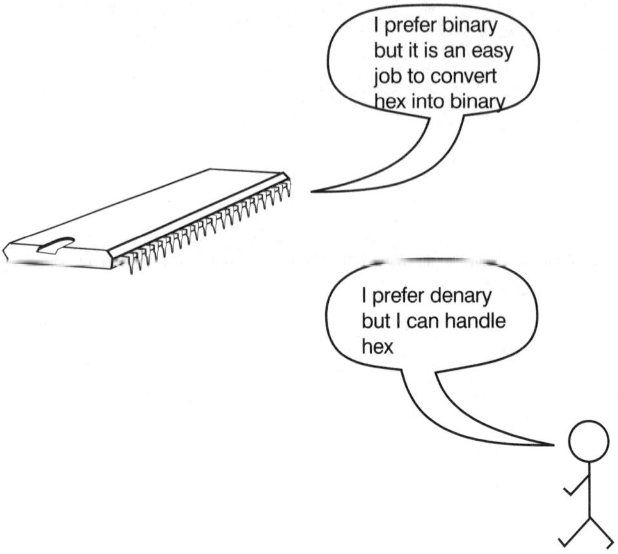

Figure 3.1

Hex is a good
compromise

Converting denary to hex

The process follows the same pattern as we saw in the denary to binary conversion.

Method

1   Write down the denary number.

2   Divide it by 16_{10}, put the whole number part of the answer underneath and the remainder in the column to the right.

3   Keep going until the number being divided reaches zero.

4   Read the answer from the bottom to top of the remainders column.

REMEMBER TO WRITE THE REMAINDERS IN HEX.

A worked example

Convert the denary number 23 823 to hex

1 Write down the number to be converted

   23 823

   (OK so far).

2 Divide by 16. You will need a calculator. The answer is 1488.9375. The 1488 can be placed under the number being converted

   23 823
   1488

but there is the problem of the decimal part. It is 0.9375 and this is actually 0.9375 of 16. Multiply 0.9375 by 16 and the result is 15. Remember that this 15 needs to be written as a hex number – in this case F. When completed, this step looks like:

```
23 823
 1488   F
```

3 Repeat the process by dividing the 1488 by 16 to give 93.0 There is no remainder so we can just enter the result as 93 with a zero in the remainder column.

```
23 823
 1488   F
   93   0
```

4 And once again, 93 divided by 16 is 5.8125. We enter the 5 under the 93 and then multiply the 0.8125 by 16 to give 13 or D in hex

```
23 823
 1488   F
   93   0
    5   D
```

5 This one is easy. Divide the 5 by 16 to get 0.3125. The answer has now reached zero and $0.3125 \times 16 = 5$. Enter the values in the normal columns to give:

```
23 823
 1488   F  =  5D0F
   93   0   ↑
    5   D   |
    0   5
```

6 Read the hex number from the bottom upwards: 5D0FH (remember that the 'H' just means a hex number).

Example

Convert $44\,256_{10}$ into hex

```
44 256
 2766   0   = ACE0H
  172   E   ↑
   10   C   |
    0   A
```

A further example

Convert $540\,709_{10}$ to hex

```
540 709
 33 794  5     = 84025H
  2112   2  ↑
   132   0  |
     8   4
     0   8
```

So $540\,709_{10} = 84025H$ but, especially when the hex number does not contain any letters, be careful to include the base of the numbers otherwise life can become really confusing.

Converting hex to denary

To do this, we can use a similar method to the one we used to change binary to denary.

Example

Convert A40E5H to denary

1 Each column increases by 16 times as we move towards the right-hand side so the column values are:

$$16^4 \quad 16^3 \quad 16^2 \quad 16^1 \quad 16^0$$
$$65536 \quad 4096 \quad 256 \quad 16 \quad 1$$

2 Simply enter the hex number using the columns

65536	4096	256	16	1
A	4	0	E	5

3 Use your calculator to find the denary value of each column

65536	4096	256	16	1
A	4	0	E	5
655360	16384	0	224	5

The left-hand column has a hex value of 10_{10} (A = 10) so the column value is $65536 \times 10 = 655360$. The next column is $4 \times 4096 = 16384$. The next column value is zero (256×0). The fourth column has a total value of $16 \times 14 = 224$ (E = 14). The last column is easy. It is just $1 \times 5 = 5$ no calculator needed!

4 Add up all the denary values:

$$655\,360 + 16\,384 + 0 + 224 + 5 = 671\,973_{10}$$

Method

1 Write down the column values using a calculator. Starting on with 16^0 (=1) on the right-hand side and increasing by 16 times in each column towards the left.

2 Enter the hex numbers in the appropriate column, converting them into denary numbers as necessary. This means, for example, that we should write 10 to replace an 'A' in the original number.

3 Multiply these denary numbers by the number at the column header to provide a column total.

4 Add all the column totals to obtain the denary equivalent.

Another example

Convert 4BF0H to denary

16^3	16^2	16^1	16^0	column values
4096	256	16	1	column values
4	11	15	0	hex values
16 384	2816	240	0	denary column totals

Total $= 16\,384 + 2816 + 240 + 0 = 19\,440_{10}$

Converting binary to hex

This is very easy. Four binary bits can have minimum and maximum values of 0000_2 up to 1111_2. Converting this into denary by putting in the column headers of: 8, 4, 2 and 1 results in a minimum value of 0 and a maximum value of 15_{10}. Doesn't this fit into hex perfectly!

This means that any group of four bits can be translated directly into a single hex digit. Just put 8, 4, 2 and 1 over the group of bits and add up the values wherever a 1 appears in the binary group.

Example

Convert 100000010101011_2 to hex

Step 1 **Starting from the right-hand end, chop the binary number into groups of four.**

100/ 0000/ 1010/ 1011/

Step 2 **Treat each group of four bits as a separate entity. The right-hand group is 1011 so this will convert to:**

8	4	2	1	column headers
1	0	1	1	binary number
8	0	2	1	column values

31

The total will then be $8 + 0 + 2 + 1 = 11_{10}$ or in hex, B.

The right-hand side binary group can now be replaced by the hex value B.

100/ 0000/ 1010/ 1011/
                          B

Step 3 **The second group can be treated in the same manner. The bits are 1010 and by comparing them with the 8, 4, 2, 1 header values this means the total value is $(8 \times 1) + (4 \times 0) + (2 \times 1) + (1 \times 1) = 8 + 0 + 2 + 0 = 10_{10}$ or in hex, A.**

We have now completed two of the groups.

100/ 0000/ 1010/ 1011/
                 A   B

Step 4 **The next group consists of all zeros so we can go straight to an answer of zero. The result so far will be:**

100/ 0000/ 1010/ 1011/
       0   A   B

Step 5 **The last group is incomplete so only the column headings of 4, 2, and 1 are used. In this case, the 4 is counted but the 2 and the 1 are ignored because of the zeros. This gives a final result of:**

100/ 0000/ 1010/ 1011/
 4   0   A   B

So, $100000010101011_2 = 40ABH$.

Having chopped up the binary number into groups of four the process is the same regardless of the length of the number. Always remember to start chopping from the right-hand side.

Example

Convert the number 1100011111001_2 to hex

Split it into groups of four starting from the right-hand side

1/ 1000/ 1111/ 1001/

Add column headers of 8 4 2 1 to each group

1	8421	8421	8421	column headings
1/	1000/	1111/	1001	binary number
1	8	8421	81	column values
1	8	15	9	group value in denary

Now just convert group values to hex as necessary. In this example only the second group 15, will need changing to F.

Final result is $1100011111001_2 = 18F9H$.

Converting hex to binary

This is just the reverse of the last process. Simply take each hex number and express it as a four bit binary number.

As we saw in the last section, a four-bit number has column header values of 8, 4, 2 and 1, so conversion is just a matter of using these values to build up the required value. All columns used are given a value of 1 in binary and all unused columns are left as zero.

When you are converting small numbers like 3H we must remember to add zeros on the left-hand side to make sure that each hex digit becomes a group of four bits.

Imagine that we would like to convert 5H to binary. Looking at the column header values of 8, 4, 2 and 1, how can we make the value 5? The answer is to add a 4 and a 1. Taking each column in turn: we do not need to use an 8 so the first column is a 0. We do want a 4 so this is selected by putting a 1 in this column, no 2 so make this 0 and finally put a 1 in the last column to select the value of 1. The 5H is converted to 0101_2. All values between 0 and FH are converted in a similar way.

Example

Convert 2F6CH to binary

Step 1 Write the whole hex number out with enough space to be able to put the binary figures underneath

$$2 \quad F \quad 6 \quad C$$

Step 2 Put the column header values below each hex digit

$$2 \quad F \quad 6 \quad C$$
$$8421 \quad 8421 \quad 8421 \quad 8421$$

Step 3 The hex C is 12_{10} that can be made of 8 + 4 so we put a binary 1 in the 8 and the 4 columns. The four-bit number is now 1100_2

$$2 \quad F \quad 6 \quad C$$
$$8421 \quad 8421 \quad 8421 \quad 8421$$
$$1100$$

Step 4 Now do the same for the next column. The hex number is 6, which is made of 4 + 2, which are the middle two columns. This will result in the binary group 0110_2

$$2 \quad F \quad 6 \quad C$$
$$8421 \quad 8421 \quad 8421 \quad 8421$$
$$0110 \quad 1100$$

Step 5 Since 8 + 4 + 2 + 1 = 15, the hex F will become 1111_2

2	F	6	C
8421	8421	8421	8421
	1111	0110	1100

Step 6 Finally, the last digit is 2 and since this corresponds to the value of the second column it will be written as 0010_2

2	F	6	C
8421	8421	8421	8421
0010	1111	0110	1100

The final result is 2F6CH = 0010111101101100_2.

But do we include the two leading zeros? There are two answers, 'yes' and 'no' but that's not very helpful. We need to ask another question: why did we do the conversion? were we doing math or microprocessors? If we were working on a microprocessor system then the resulting 16 bits would represent 16 voltages being carried on 16 wires. As the numbers change, all the wires must be able to switch between 0 V and 3.3 V for binary levels 0 and 1. This means, of course, that all 16 wires must present so we should include the binary levels on all of them.

If the conversion was purely mathematical, then since leading (left-hand end) zeros have no mathematical value there is no point in including them in the answer.

Method

1 Write down the hex number but make it well spaced.

2 Using the column header values of 8, 4, 2 and 1, convert each hex number to a four bit binary number.

3 Add leading zeros to ensure that every hex digit is represented by four bits.

Example

Convert 1E08BH to binary

Step 1

1	E	0	8	B
8421	8421	8421	8421	8421

Step 2

0001	1110	0000	1000	1011

So, 1E08BH = 00011110000010001011_2.

Using stepping stones

It is fairly easy to convert binary to hex and hex to binary. I find it much easier to multiply and divide by 2 rather than by 16, so when faced with changing hex into denary and denary into hex I often change them into binary first. It is a longer route but at least I can do it without my calculator (see Figure 3.2).

Figure 3.2

A longer route may prove easier

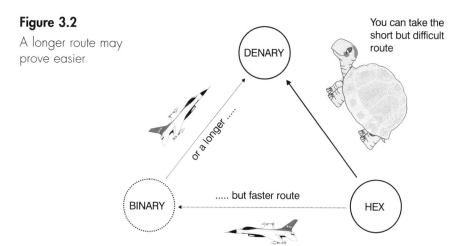

Obsolete octal – probably not worth reading

Octal is another number system which has no advantages over hex but is still met from time to time. Only a brief look will be offered here just to make sure that we have at least mentioned it.

In hex, we used binary bits in groups of four because 1111_2 adds up to 15 which is the value of the highest digit (F) in hex. In octal, we use groups of three bits. The highest value is now 111_2 which is 7. Octal therefore has eight digits and counts from 0 to 7. The count proceeds:

```
0
1
2
3
4
5
6
7
```

There is no 8th digit so reset the count to 0 and put a 1 in the next column.

10
11
12
13
14
15
16
17

Now go straight to 20

20
etc.

No letters are involved and it is often not recognized as octal until we realize that none of the numbers involve the digits 8 or 9.

Conversions follow the same patterns as we have seen for hex.

Octal to denary: the column heading values are 8^4, 8^3, 8^2, 8^1, 8^0.

Denary to octal: divide by 8 and write down the remainder then read remainders from the bottom upwards. Use the subscript 8 to indicate an octal number, e.g. $64_{10} = 100_8$.

Octal to binary: write each octal digit down as a *three* digit binary group.

Binary to octal: start from the right-hand side and chop the binary numbers into groups of three, then evaluate each group.

I think that is enough for octal. It's (fairly) unlikely you will meet it again so we can say 'goodbye Octal'.

Quiz time 3

In each case, choose the best option.

1   Which of these represents the largest number?

(a) 1000_8
(b) 1000_{10}
(c) 1000_2
(d) 1000H

2   The number CD02H is equal to:

(a) 52482_{10}
(b) 54228_{10}
(c) 56322_{10}
(d) 52842_{10}

3   The base of a number system is:

(a) always the same as the highest digit used in the system.
(b) usually +5 or +3.3.
(c) equal to the number of different digits used in the system.
(d) one less than the highest single digit number in the system.

**4   Which of these numbers is the same as
101101110102:**

(a) 1646_{10}
(b) $5BA_{16}$
(c) AB5H
(d) B72h

5   The number of digits in a denary number is often:

(a) more than the number of digits in the equivalent binary number.
(b) less than or equal to the number of digits in the equivalent hex number.
(c) more than the number of digits in the equivalent hex number.
(d) more than the number of digits in the equivalent decimal number.

4

How micros calculate

How the microprocessor handles numbers (and letters)

In the last chapter, we saw how numbers could be represented in binary and hex forms. Whether we think of a number as hex or binary or indeed denary, inside the microprocessor it is only binary. The whole concept of hex is just to make life easier for us.

We may sit at a keyboard and enter a hex (or denary) number but the first job of any microprocessor-based system is to convert it to binary. All the arithmetic is done in binary and its last job is to convert it back to hex (or denary) just to keep us smiling.

There was a time when we had to enter binary and get raw binary answers but thankfully, those times have gone. Everything was definitely NOT better in the 'good old days'.

The form binary numbers take inside of the microprocessor depends on the system design and the work of the software programmers. We will take a look at the alternatives, starting with negative numbers.

In real life, it is easy, we just put a – symbol in front of the number and it is negative so +4 becomes –4. Easy, but we don't have any way of putting a minus sign inside the microprocessor. We have tried several ways round the problem.

Signed magnitude numbers

The first attempt seemed easy but it was false optimism. All we had to do was to use the first bit (msb) of the number to indicate the sign 1 = minus, 0 = plus.

This had two drawbacks.

1 It used up one of the bits so an 8-bit word could now only hold seven bits to represent numbers and one bit to say 'plus' or 'minus'. The seven bits can now only count up to $1111111_2 = 127$ whereas the eight bits should count to 255.
2 If we added two binary numbers like +127 and +2, we would get:

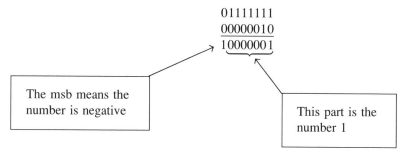

The msb means the number is negative

This part is the number 1

The msb (most significant bit) of 1 means it is a minus number and the actual number is 0000001 = 1. So the final result of +127 + 2 is not 129 but minus 1.

When we use a microprocessor to handle arithmetic with these problems, we can ensure that the microprocessor can recognize this type of accidental negative number. We can arrange for the microprocessor to compensate for it but it is rather complicated and slow.

Luckily, a better system came along which has stood the test of time, having been used for many years.

Complementary numbers

This has two significant advantages:

1 It allows the full number of bits to be used for a number so an 8-bit word can count from 0 to 11111111_2 or 255.
2 It is easy to implement with addition and subtraction using substantially the same circuitry.

So, how do we manage to use all eight bits for numbers yet still be able to designate a number positive or negative?

That's clever. We will start by looking at positive numbers first because it is so easy. All positive numbers from 0 to 255 are the same as we get by simply converting denary to binary numbers. So that's done.

39

Addition

Example

Add 01011010 + 00011011.

The steps are just the same as in 'normal' denary arithmetic.

Step 1 Lay them out and start from the lsb (least significant bit) or right-hand bit

```
0 1 0 1 1 0 1 0 +
0 0 0 1 1 0 1 1
```

Add the right-hand column and we have 0 + 1 = 1.

So we have

```
0 1 0 1 1 0 1 0 +
0 0 0 1 1 0 1 1
              1
```

Step 2 Next we add the two 1s in the next column. This results in 2, or 10 in binary. Put the 0 in the answer box and carry the 1 forward to the next column

```
0 1 0 1 1 0 1 0 +
0 0 0 1 1 0 1 1
            0 1
          1
```

Step 3 The next column is easy 0 + 0 + 1 = 1

```
0 1 0 1 1 0 1 0 +
0 0 0 1 1 0 1 1
          1 0 1
          1
```

Step 4 The next line is like the second column, 1 + 1 = 10. This is written as an answer of 0 and the 1 is carried forward to the next column

```
0 1 0 1 1 0 1 0 +
0 0 0 1 1 0 1 1
        0 1 0 1
        1   1
```

Step 5 We now have 1 in each row and a 1 carried forward so the next column is 1 + 1 + 1 = 3 or 11 in binary. This is an answer of 1 and a 1 carried forward to the next column

```
0 1 0 1 1 0 1 0 +
0 0 0 1 1 0 1 1
      1 0 1 0 1
    1 1   1
```

Step 6   **The next column is 0 + 0 + 1 = 1, and the next is 1 + 0 = 1 and the final bit or msb is 0 + 0 = 0, so we can complete the sum**

```
0 1 0 1 1 0 1 0 +
0 0 0 1 1 0 1 1
0 1 1 1 0 1 0 1 ──────  90 + 27 = 117
  1 1     1
```

Subtraction

Here is a question to think about: What number could we add to 50 to give an answer of 27? In mathematical terms this would be written as $50 + x = 27$.

What number could x represent? Surely, anything we add to 50 must make the number larger unless it is a negative number like -23:

$$50 + (-23) = 27$$

The amazing thing is that there is a number that can have the same effect as a negative number, even though it has no minus sign in front of it. It is called a 'two's complement' number.

Our sum now becomes:

50 + (the two's complement of 23) = 27

This magic number is the *two's complement* of 23 and finding it is very simple.

How to find the two's complement of any binary number

Invert each bit, then add 1 to the answer

All we have to do is to take the number we want to subtract (in its binary form) and invert each bit so every one becomes a zero and each zero becomes a one. Note: technically the result of this inversion is called the 'one's complement' of 23. The mechanics of doing it will be discussed in the next chapter but it is very simple and the facility is built into all microprocessors at virtually zero cost.

Converting the 23 into a binary number gives the result of 00010111_2 (using eight bits). Then invert each bit to give the number 11101000_2 then add 1. The resulting number is then referred to as the 'two's complement' of 23.

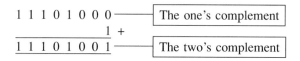

```
1 1 1 0 1 0 0 0 ──────  The one's complement
            1 +
1 1 1 0 1 0 0 1 ──────  The two's complement
```

41

In this example, we used 8-bit numbers but the arithmetic would be exactly the same with 16 bits or indeed 32 or 64 bits or any other number.

Doing the sum

We now simply add the 50 and the two's complement of 23:

50 + (the two's complement of 23) = 27

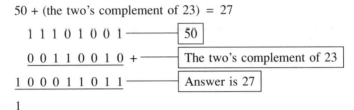

The answer is 100011011.

Count the bits. There are nine! We have had a carry in the last column that has created a ninth column. Inside the microprocessor, there is only space for eight bits so the ninth one is not used. If we were to ask the microprocessor for the answer to this addition, it would only give us the 8-bit answer: 00011011_2 or in denary, 27. We've done it! We've got the right answer!

It was quite a struggle so let's make a quick summary of what we did.

1 Convert both numbers to binary.
2 Find the two's complement of the number you are taking away.
3 Add the two numbers.
4 Delete the msb of the answer.

Done.

A few reminders

1 Only find the two's complement of the number you are taking away – NOT both numbers.
2 If you have done the arithmetic correctly, the answer will always have an extra column to be deleted.
3 If the numbers do not have the same number of bits, add leading zeros as necessary as a first job. Don't leave until later. Both of the numbers must have the same number of bits. They can be 8-bit numbers as we used, or 16, or 32 or anything else so long as they are equal.

A quick way to find the two's complement of a binary number

Start from the left-hand end and invert each bit until you come to the last figure 1. Don't invert this figure and don't invert anything after it.

Example 1

What is -24_{10} expressed as an 8-bit two's complement binary number?

1 Change the 24_{10} into binary. This will be 11000.
2 Add leading zeros to make it an 8-bit number. This is now 00011000.
3 Now start inverting each bit, working from the left until we come to the last figure '1'. Don't invert it, and don't invert the three zeros that follow it.

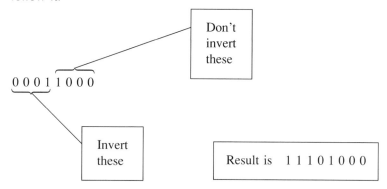

0 0 0 1 1 0 0 0

Don't invert these

Invert these

Result is   1 1 1 0 1 0 0 0

Example 2

What is -100_{10} expressed as a 16-bit two's complement binary number?

1 Convert the 100_{10} into binary. This gives 1100100_2.
2 Add nine leading zeros to make the result the 16-bit number 0000000001100100.
3 Now, using the quick method, find the two's complement:

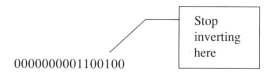

0000000001100100

Stop inverting here

The result is   1111 1111 1001 1100

Example 3

Find the value of $1011\ 0111_2 - 00\ 1011_2$ using two's complement addition.

1 The second number has only six bits so add two zeros on the left-hand end to give 1011 0111 – 0000 1011.
2 Invert each bit in the number to be subtracted to find the one's complement. This changes the 00001011 to 11110100.
3 Add 1 to give the two's complement: 11110100 + 1 = 11110101 (or do it the quick way).

4 Add the first number to the two's complement of the second number:

```
  1 0 1 1 0 1 1 1
  1 1 1 0 1 0 1  +
1 1 0 1 0 1 1 0 0
1 1 1 1   1 1 1
```

5 The result so far is 110101100 which includes that extra carry so we cross off the msb to give the final answer of 10101100_2.

Floating point numbers

Eight-bit numbers are limited to a maximum value of 11111111_2 or 255_{10}. So, 0 – 255 means a total of 256 different numbers. Not very many. 32-bit numbers can manage about 4¼ billion. This is quite enough for everyday work, though Bill Gates' bank manager may still find it limiting. The problem is that scientific studies involve extremely large numbers as found in astronomy and very small distances as in nuclear physics.

So how do we cater for these? We could wait around for a 128-bit microprocessor, and then wait for a 256-bit microprocessor and so on. No, really, the favourite option is to have a look at alternative ways of handling a wide range of numbers. Rather than write a number like 100 we could write it as 1×10^2. Written this way it indicates that the number is 1 followed by two zeros and so a billion would be written as 1×10^9. In a similar way, 0.001 is a 1 preceded by two zeros would be written as 1×10^{-3} and a billionth, 0.000000001, would be 1×10^{-9}. The negative power of ten is one greater than the number of zeros. By using floating point numbers, we can easily go up to 1×10^{99} or down to 1×10^{-99} without greatly increasing the number of digits.

Fancy names

Normalizing

Changing a number from the everyday version like 275 to 2.75×10^2 is called normalizing the number. The first number always starts with a single digit between 1 and 9 followed by a power of ten.

In binary we do the same thing except the decimal point is now called a binary point and the first number is always 1 followed by a power of two as necessary.

Three examples

1 Using the same figure of 275, this could be converted to 100010011 in binary. This number is normalized to 1.00010011×2^8.
2 A number like 0.0001001_2 will have its binary point moved four places to the *right* to put the binary point just after the first figure 1 so the normalized number can be written as 1.001×2^{-4}.

3 The number 1.101_2 is already normalized so the binary point does not need to be moved so, in a formal way, it would be written as 1.101×2^0.

A useless fact

Anything with a power of zero is equal to 1. So $2^0 = 1$, $10^0 = 1$. It is tempting but total nonsense to use this fact to argue that since $2^0 = 1$ and $10^0 = 1$ then 2 must equal 10!

Terminology

There are some more fancy names given to the parts of the number to make them really scary.

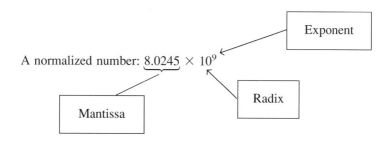

The exponent is the power of ten, in this example, 9. The mantissa, or magnitude, is the number, in this case 8.0245. The radix is the base of the number system being used, 2 for binary, 16 for hex, 10 for decimal.

Storing floating point numbers

In a microprocessor, the floating point is a binary number. Now, in the case of a binary number, the mantissa always starts with 1 followed by the binary point. For example, a five digit binary mantissa would be between 1.0000 and 1.1111.

Since all mantissas in a binary system start with the number 1 and the binary point, we can save storage space by missing them out and just assuming their presence. The range above would now be stored as 0000 to 1111.

It is usual to use a 32-bit storage area for a floating point number. How these 32 bits are organized is not standardized so be careful to check before making any assumptions. Within those 32 bits, we have to include the exponent and the mantissa which can both be positive or negative. One of the more popular methods is outlined below.

45

Exponent	Mantissa	S

Bit   31   24   23   1   0

Bit 0 is used to hold the sign-bit for the mantissa using the normal convention of 0 = positive and 1 = negative.

Bits 1–23 hold the mantissa in normal binary.

Bits 24–31 hold the exponent. The eight digits are used to represent numbers from –127 to +128 using either two's complement numbers or excess-127 notation.

We have already met two's complement numbers earlier in this chapter so we will look at excess-127 notation now.

Excess-127 notation

This is very simple, despite its impressive name. To find the exponent just add 127 to its value then convert the result to binary. This addition will ensure that all exponents have values between 0 and 255, i.e. all positive values.

Example

If the exponent is –35 then we add 127 to give the result 92, which we can then convert to binary (01011100).

When the value is to be taken out of storage and converted back to a binary number, the above process is reversed by subtracting the 127 from the exponent.

Size, accuracy and speed

The mantissa can go as high as $1.1111\ 1111\ 1111\ 1111\ 1111\ 111_2$. To the right of the binary point the decimal equivalents are values of 1.5 + 0.25 + 0.125 + 0.0625 etc. Adding these up gives a total that is virtually 2 – but not quite. The larger the number of bits in the mantissa, the more accuracy we can expect in the result. The exponent has eight bits so it can range from –127 to +128 giving a maximum number of 1×2^{128} which is approximately 3.4×10^{38}. The accuracy is limited by the number of bits that can be stored in the mantissa, which in this case is 23 bits.

If we want to keep to a total of 32 bits, then we have a trade-off to consider. Any increase in the size of the exponent, to give us larger numbers, must be matched by reducing the number of bits in the mantissa that would have the effect of reducing the accuracy. Floating point operations per second (FLOPS) is one of the choices for measuring speed.

IBM are building (2002) a new super computer employing a million microprocessors. The Blue Gene project will result in a computer running at a speed of over a thousand million million operations per second (1 petaflop). This is a thousand times faster that the Intel 1998 world speed record or about two million times faster than the current top-of-the-range desktop computers.

Single and double precision

If we need more accuracy, an alternative method is to increase the number of bits that can be used to store the number from 32 (single-precision) to 64 (double-precision). If this extra storage space is devoted to increasing the mantissa bits, then the accuracy is increased significantly.

Binary coded decimal (BCD)

Binary coded decimal numbers are very simple. Each decimal digit is converted to binary and written as a 4-bit or 8-bit binary number. The number 5 would be written as 0101_2 or 00000101_2. So far, this is the same as 'ordinary' binary but the change occurs when we have more digits.

Consider the number 25_{10}. In regular binary this would convert to 11001_2. Alternatively, we could convert each digit separately to 4-bit or 8-bit numbers:

$2 = 0010_2$ or $0000\ 0010_2$
$5 = 0101_2$ or $0000\ 0101_2$

Putting these together, 25_{10} could be written using the 4-bit numbers as $0010\ 0101_2$. This uses one byte and is called Packed BCD.

Alternatively, we could use the 8-bit formats and express 25_{10} as $0000\ 0010\ 0000\ 0101_2$ and would now use two bytes. This is called Unpacked BCD.

There are two disadvantages. Firstly, many numbers are of increased length after converting to BCD, particularly so if we use unpacked BCD or the numbers are very large like 25×10^{75}. In addition, arithmetic is much more difficult although, generally, microprocessors do have the ability to handle them.

The advantage becomes apparent when the microprocessor is controlling an external device like digits on displays at a filling station or accepting inputs from a keyboard. The coding is simple and does not involve the conversion of the numbers to binary.

47

> **Overall**
> Arithmetic → use binary
> Inputting and outputting numbers → use BCD

Quiz time 4

In each case, choose the best option

1 **The number -35_{10}, when expressed as an 8-bit binary number in two's complement form, is:**

(a) 00100011.
(b) 1111011101.
(c) 11011101.
(d) 00110101.

2 **The number 7_{10} converted to an unpacked BCD format would be written as:**

(a) 1110 0000.
(b) 7H.
(c) 0000 0111.
(d) 0111.

3 **The signed magnitude number 11001100_2 is equivalent to:**

(a) -76_{10}.
(b) 204_{10}.
(c) CCH.
(d) 1212_{10}.

4 **In the number 0.5×10^{24} the number:**

(a) 10 is the mantissa.
(b) 24 is the exponent.
(c) 0 is the sign bit.
(d) 5 is the radix.

5 **A signed magnitude number that has a figure:**

(a) zero as the msb is a negative number.
(b) one as the lsb is a negative number.
(c) one as the msb is a negative number.
(d) zero as the lsb is a negative number.

5

An introduction to logic gates and their uses

Opening and closing gates

In the last chapter the binary values zero and one are represented by two different voltages. Binary zero is a voltage close to 0 V and binary one by a voltage close to +5 V (some logic circuits use other voltage levels but this is a popular value and will serve as an example).

A gate is a simple electronic circuit that has a single output voltage that corresponds to one of the two binary values. These gates are often referred to as 'logic gates' and the output voltages as 'logic 0' or 'logic 1' instead of binary 0 and 1. The distinction is just in the name. If you were to ask a mathematician or a computer programmer, they will refer to the outputs as binary values but an electronics engineer will call them logic levels. It really doesn't matter.

What decides the output voltage?

We connect one or more voltages to the input of the gate. These input voltages are either logic 0 or logic 1 levels. The logic gate looks at the input voltages and 'decides', depending on its design, what voltage to produce at the output of the circuit.

There are only four basic designs of gate. They are called the NOT gate, the AND gate, the OR gate and the XOR gate. Notice how we use capital letters for the names of the gates otherwise we can finish up with some almost indecipherable sentences. Not not or and not and or not . . .

A little reminder before we start. Logic gates are clever little chaps but they are not magic. Just like any other electronic circuit, they need power supplies to make them work. Now, because all gates and microprocessors need power supplies, we tend to assume that everyone knows that. You will notice that power supplies are not shown in any of the diagrams in this chapter but that doesn't mean that they are not there!

We will explore these gates now, starting from the simplest.

The NOT gate

It has only one input and performs a very simple function. It simply reverses the binary value. If we put a logic 1 into it, we get a logic 0 at the output. Similarly, a 0 at the input gives a 1 at the output. On a diagram, we represent a NOT gate by a symbol as shown in Figure 5.1.

Figure 5.1

Symbols for a NOT gate

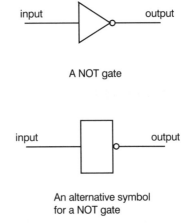

A NOT gate

An alternative symbol
for a NOT gate

A truth table

This is an alternative to the wordy description of how a gate works. It simply lists all the possible inputs to the gate together with the corresponding outputs. The truth table for a NOT gate is really easy. There are only two possible inputs: 0 and 1 as we can see in Figure 5.2.

So, how is it used in the microprocessor?

The truth table only shows what happens to a single bit but in the microprocessor we may want to use a NOT gate to invert a hex number like A4H. In this case the hex number is first converted to an 8-bit binary number. This process is not performed by the

Figure 5.2

The truth table for a NOT gate

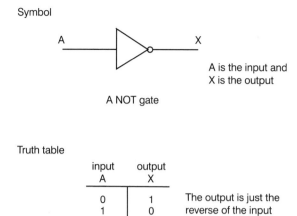

Symbol

A ———▷o——— X

A is the input and
X is the output

A NOT gate

Truth table

input A	output X	
0	1	The output is just the
1	0	reverse of the input

microprocessor but by other external circuits. By the time it reaches the microprocessor it has been converted to the binary equivalent of 10100100_2.

The NOT gate has only one input so, to handle an 8-bit binary word, we will need eight NOT gates. Now it becomes much easier. Each NOT gate inverts just one of the bits and all the outputs are grouped together to form a new hex number. See how it works in Figure 5.3. The result was the hex number 5BH. This is curious. If we add A4H to this result of 5BH we get FFH or all 'ones' in binary, 11111111_2. There was nothing special about the number A4H. It happens with any pair of numbers generated by NOT gates. Why is this? Figure 5.3 gives a clue.

Figure 5.3

Inverting a hex number

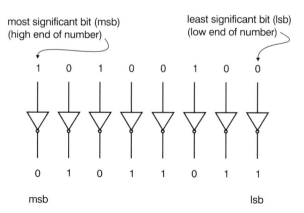

Apply the binary word to eight NOT gates

most significant bit (msb)
(high end of number)

least significant bit (lsb)
(low end of number)

1 0 1 0 0 1 0 0

0 1 0 1 1 0 1 1

msb   lsb

Now converting the binary to hex gives 5BH

A little extra bit

We can show an inversion by drawing a line over the top. In Figure 5.2 the input was given the letter A and the output was shown as X. We could say: X = Ā.

AND gate

Unlike the NOT gate, an AND gate has more than one input. In fact we can have as many inputs as we like but the good news is that in microprocessors only two inputs are used. This simplifies the symbols and the truth tables considerably.

An AND gate is any circuit that gives a logic (or binary) 1 if (and only if) every input to the circuit is at logic 1. So in microprocessors, with only two inputs, it is easier to say that it gives a 1 out if both of the inputs is at logic 1. The symbols for the AND gate are shown in Figure 5.4.

Figure 5.4

Symbols for an AND gate

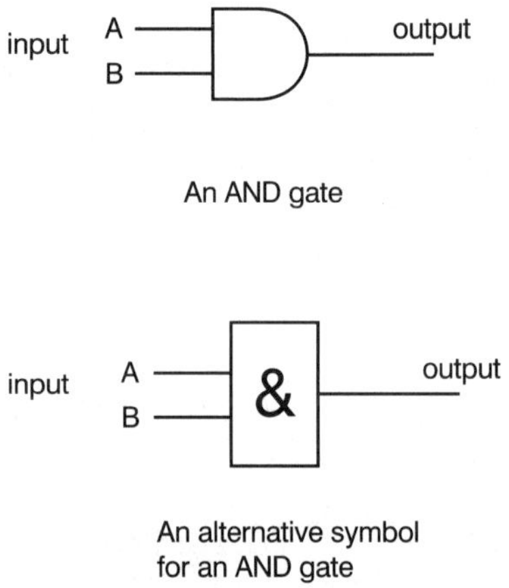

input A
      B
output

An AND gate

input A & output
      B

An alternative symbol
for an AND gate

The truth table in Figure 5.5 has four rows to cover all the possible combinations of inputs.

What is the point of an AND gate?

We often meet an AND gate without realizing it. When we climb into an elevator the door must be closed AND the floor button pressed before the motor will start. This is an AND gate in action.

In a microprocessor, groups of AND gates are used to handle pairs of inputs at the same time just like we did with the NOT gates. For convenience we usually use hex numbers to describe groups of

Figure 5.5

The truth table for an AND gate

Symbol

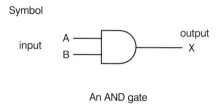

An AND gate

Truth table

inputs		output
A	B	X
0	0	0
0	1	0
1	0	0
1	1	1

The output is always 0 unless both inputs are 1

inputs and outputs but remember it's all really happening in binary.

Unless we appreciate that the hex numbers are really just groups of ones and zeros, it may seem odd to talk about putting numbers through AND gates. Figure 5.6 may make it (a little) clearer. In this figure, we have used hex inputs of 37H and 5BH giving a result of 13H. Note that the AND gate does not add numbers.

A little extra bit

In data books and manuals, the AND function is abbreviated to a dot like a period (full-stop). So, if an AND gate had two inputs called A and B and an output called X, then we could write X = A.B

Sometimes it is further simplified to X = AB

Very occasionally we come across a symbol $\wedge$ so we can write X = A $\wedge$ B.

Figure 5.6

An AND gate in action

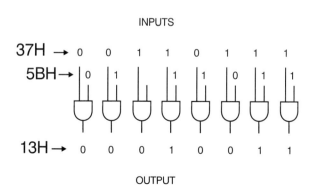

INPUTS

OUTPUT

The NAND gate

The word NAND is just a fancy contraction of NOT and AND. The NAND gate is just an AND gate followed by a NOT gate so all the outputs shown in the AND truth table are just inverted by the NOT gate.

On a diagram, this combination is indicated by putting a small circle on the end of the symbol. The symbol and truth table are shown in Figure 5.7.

Figure 5.7

The truth table for a NAND gate

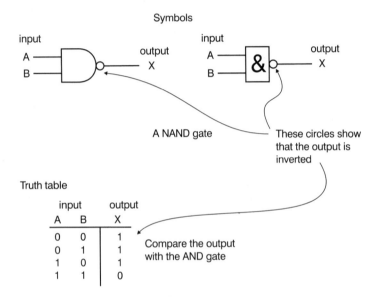

It is better not to make a big effort to learn the NAND gate. Just remember that it's the same as the AND gate except the output has been inverted by the NOT gate that has been added internally.

The symbol has a line over the top to indicate the added NOT function. A two input NAND could be written as $X = \overline{A.B}$ or $X = \overline{AB}$ or very occasionally $X = A \wedge B$.

The OR gate

This follows on nicely from the AND gate. The OR gate gives a logic one at its output if either (or both) of the inputs is at a logic one. Just like the AND gate, the OR gate can have as many inputs as we wish but in a microprocessor, we only use two input versions. Its symbols and truth table are shown in Figures 5.8 and 5.9.

54

Figure 5.8

Symbols for an OR gate

An OR gate

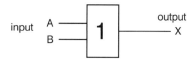

An alternative symbol
for an OR gate

Figure 5.9

The truth table for an
OR gate

An OR gate

Truth table

input		output
A	B	X
0	0	0
0	1	1
1	0	1
1	1	1

The output is always 1
unless both inputs are 0

Another extra bit

The OR function can be written as + or sometimes $\vee$. So, if an OR gate had two inputs called A and B and an output called X, then we could write $X = A + B$ or $X = A \vee B$. Don't mistake this + sign as 'plus' as in addition $3 + 4 = 7$.

The NOR gate

As we would expect, this is just the same as the OR gate except for the NOT gate added to the output. The symbol has the inversion line over it to give $X = \overline{A + B}$ or $\overline{A \vee B}$ (see Figure 5.10).

55

Figure 5.10

The NOR gate

  Symbols

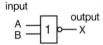

A NOR gate

Truth table

input A	B	output X	
0	0	1	This is just the opposite
0	1	0	to the OR gate output
1	0	0	
1	1	0	

The XOR gate

This is called the Exclusive-OR gate, which is abbreviated to XOR or EOR. Here are two examples of everyday English, both using the word 'or'.

We get into an elevator and the operator says to us 'Do you want to go up or down?' We have a choice, we can decide to go up or we can go down. We have two possible answers – 'up' or 'down'.

On the way home we buy a burger. We are asked 'Do you want ketchup or mustard?' This time we could answer 'ketchup' or 'mustard' or we could say 'both'.

Figure 5.11

The XOR or difference gate

Symbol

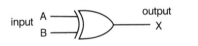 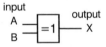

An XOR gate

Truth table

input A	B	output X	
0	0	0	This is an inverted
0	1	1	version of the XOR
1	0	1	
1	1	0	

This is a fine example of how we can make something which is really easy appear difficult. The word 'or' in English has two different meanings which can be referred to as exclusive and inclusive. The first example used the exclusive 'or' because we could have one or the other *but not both*. The second situation uses the inclusive 'or' because we could have one or the other or both. We use both meanings everyday and understand them so well that we know automatically which one is meant.

In the XOR gate the output is a logic one if either input is a logic 1 *but not both*. Only two-input XOR gates are manufactured, whether for use in a microprocessor or not (see Figure 5.11).

To look at it in another way, the output from an XOR is a logic one if the two inputs are different. For this reason, it is often called a 'Difference' gate.

The electronic padlock

When we want to take some money out of our bank account we can go to the money machine and pop our card in. It then asks us to key in our code number which is usually a four-digit number. If the number is correct, we have access to our account.

In the background is a microprocessor with two useful attributes:

1 Microprocessors are very good at spotting whether a number is zero or not.
2 They contain a series of XOR gates.

So how can it check my number? One easy method is to compare our keyed-in number with the number read from the magnetic strip on the back of the card. Each of our code numbers is treated as a hex number and converted to a four-bit binary number. This results in one 16-bit binary number from the keyboard and another from the magnetic strip on the card, which now have to be compared.

Inside the microprocessor are 16 XOR gates each with two inputs – one from the card and one from the keyboard. Every time the bits coincide, whether they are both zeros or both ones, the output will be a zero.

The 16 results are quickly scanned looking for any output which is not zero. This would indicate an incorrect number. The process is shown in Figure 5.12 but for clarity, only four of the XOR gates have been shown. In real life, there will be 16 of them, of course.

The extra bit

The XOR function can be written as $\oplus$. So, if an XOR gate had inputs called A and B and an output called X, then it would be abbreviated as $X = A \oplus B$.

Figure 5.12

Using an XOR gate to compare numbers

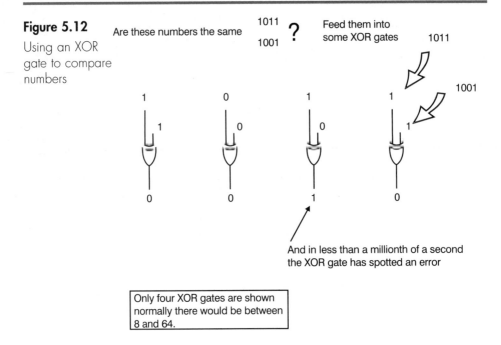

Are these numbers the same 1011 **?** Feed them into some XOR gates 1011

1001

1011

1001

And in less than a millionth of a second the XOR gate has spotted an error

Only four XOR gates are shown normally there would be between 8 and 64.

The XNOR (or ENOR) gate

This is the inverted version of the XOR gate. These result in the output being at logic 1 only when the two inputs have the same value or are equivalent. For this reason, it is often referred to as the Equivalence gate (see Figure 5.13).

Figure 5.13

The truth table for an XNOR or equivalence gate

Symbol

input A
B
output X

input A
B
$=1$
output X

An XNOR gate

Truth table

| input | | output |
A	B	X
0	0	1
0	1	0
1	0	0
1	1	1

This is an inverted version of the XOR

The tri-state buffer

This looks like a logic gate but behaves more like a switch. In Figure 5.14 we can see that it is quite simple having only an input, an output and another connection called an 'enable'. The purpose of the enable line is to switch the buffer on or off. When the buffer is switched on, any signal applied to the input appears at the output and when it is switched off, the buffer is disconnected so that there is no output signal present.

Figure 5.14

A tri-state buffer is like a switch

Symbol

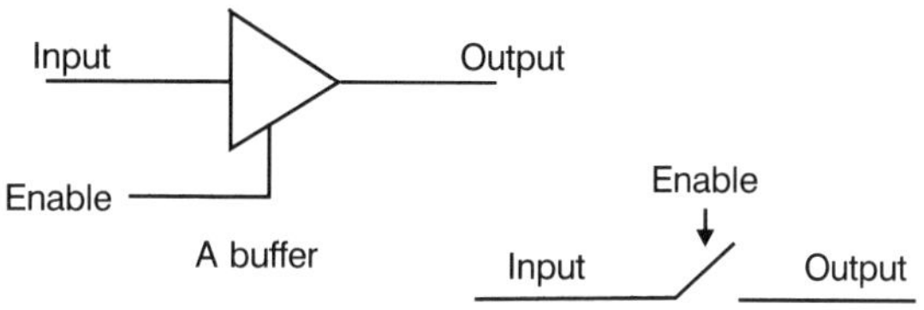

Input — Output

Enable

A buffer

Enable

Input — Output

Truth table

input	output
0	0
1	1

The output is just the same as the input (or not there at all!)

So, why not just use a switch?

The problem with a switch is that, once closed, the input and output are physically joined so that input and output circuits are connected together. The buffer is a one-way device for signals so that the output is isolated from the input to prevent any changes in the next circuit from interfering with the input circuits.

Look for the circles

In the NOT, NAND NOR and XNOR gates a small circle was shown at the output to indicate that the output has been inverted. The same

Figure 5.15

An active-low tri-state buffer

The required voltage on the enable can be changed.

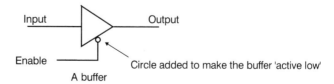

Input — Output

Enable

A buffer

Circle added to make the buffer 'active low'

Active low means the enable must be at logic 0 to switch the buffer ON and allow the data to pass through

Fiqure 5.16

The four types of tri-
state buffer

Enable = 1
output is
not inverted

Enable = 1
output is
inverted

Enable = 0
output is
not inverted

Enable = 0
output is
inverted

thing occurs with the buffer. If a circle is shown, the output is an
inverted copy of the input.

We apply a similar convention to the enable input. If the input has to
be a logic 1 level to switch it on, then it is as shown in Figure 5.15. If,
however, the enable input has to be a logic 0 to enable it a small circle
is shown at the point where the enable line connects with the buffer.
When the buffer is switched off, it is said to be disabled. This would
give us four possible buffers as in Figure 5.16.

Quiz time 5

In each case, choose the best option.

---

1   **Which of the gates shown in Figure 5.17 would have
    an output of logic zero?**

    (a) A, B, C.
    (b) A and D.
    (c) B and D.
    (d) None of them.

Figure 5.17

Which outputs are
at logic 0?

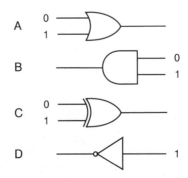

2   If the hex number 4C is applied to eight NOT gates, the output would have the value of:

(a) B3.
(b) 4C.
(c) FF.
(d) CD.

3   Which two-input gate has an output of logic 0 when both inputs are at logic 1:

(a) OR.
(b) AND.
(c) NOT.
(d) XOR.

4   A gate is:

(a) necessary to keep the cattle in the field.
(b) always operated from a supply of +5 V.
(c) an electronic circuit with three connecting wires.
(d) an electronic switching circuit whose output voltage depends on the inputs.

5   Adding the input and output binary values of a NOT gate:

(a) can be 0, 1 or 2 depending on the inputs.
(b) will give a voltage around 2.5 V.
(c) always results in 1.
(d) is not possible.

6

Registers and memories

The logic gates that we met in the last chapter occur in their millions in microprocessors and in the surrounding circuitry. They are to be found in all microprocessors from the oldest and simplest, to this years' 'Ultimate Wonder Child' and even next year's 'New and Improved Ultimate Wonder Child MkII'.

When logic gates are used in a microprocessor, they are usually grouped together into circuits, called flip-flops, each one being able to store a single binary digit.

A flip-flop or bistable

A flip-flop or bistable is a circuit that can store a single binary bit – either 0 or 1. One useful characteristic of the flip-flop is that it can only have an output of 0 or 1. It cannot hover somewhere in between. The flip-flop is shown in Figure 6.1. The purpose of the clock input is to tell the flip-flop when to accept the new input level.

The sequence of the events is:

1 Apply the binary level to be stored.
2 Wait a short time (a few nanoseconds) until the voltage is properly established.
3 Apply a signal to the clock input to tell the flip-flop to memorize the signal present at the input.

Figure 6.1

A flip-flop – the basic building block of a microprocessor

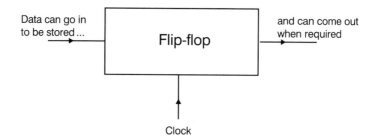

Data can go in to be stored ...

Flip-flop

and can come out when required

Clock

Why do we have to wait?

When we apply a voltage to a length of wire, we would hope that the voltage changes as in Figure 6.2. Unfortunately, it takes a few nanoseconds to settle down. The rise of voltage travels along the connecting wire and is reflected from the end causing another voltage to be reflected towards the input. This reflection is itself reflected and after repeated reflections, the voltage slowly settles down to its new level as in Figure 6.3. This occurs if we suddenly tip a bucket of water into a half-filled bath. The added water sets up a wave that is reflected

Figure 6.2

A voltage is switched on – how we would like it to change

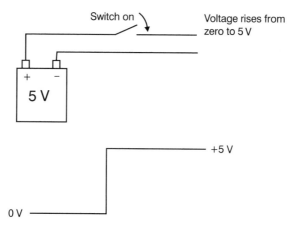

Switch on

Voltage rises from zero to 5 V

+ –
5 V

+5 V

0 V

Figure 6.3

A voltage is switched on – what really happens?

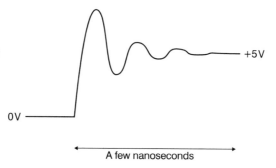

+5V

0V

A few nanoseconds

backwards and forwards along the bath as the new level is established. If we didn't wait for the voltage to settle down, we could accidentally store an incorrect value.

And what about the clock signal?

This is just an input to tell the flip-flop that it is time to read the input level. All microprocessor operations are carefully timed by clock pulses to ensure that the system operates in the correct sequence.

The clock signal is usually a positive-going voltage pulse. This pulse can be used to switch two circuits at different times by designing one circuit to respond to an increasing voltage and the other to use a decreasing voltage. If, for example, the pulse in Figure 6.4 were to be 10 ns wide then this could create the required delay for the voltage to settle. The circuit supplying the input voltage is kicked into action by the positive-going edge and then 10 ns later the negative-going edge instructs the flip-flop to save the data present at that time.

Figure 6.4

Using a clock pulse to control timing of a circuit

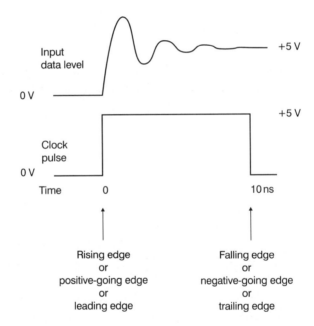

A register

A register is just a collection of flip-flops. A flip-flop can only store one bit so to handle 32 bits at a time we would need 32 flip-flops and would refer to this as a 32-bit register. To save space, Figure 6.5 shows an 8-bit register.

The register has two distinct groups of connections: the data bits 0 to 7 and the control signals. The data connections or data lines carry the

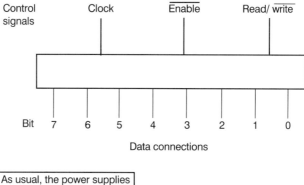

Figure 6.5

An 8-bit register

binary levels in or out of the register. The number of data lines determines the size of the register so a 64-bit register would have 64 data connections.

The three control signals include two new ones

1 Enable. This is a simple on/off switch for the register. We met this in Chapter 5 with the tri-state buffer. The line over the top of the word indicates that it is 'on' when this line is 'low' or at logic zero. We tend to say the line is 'active low' in this situation. Therefore, it follows that the register is disabled or switched 'off' when the enable line is at logic 1 or 'high'. Nearly all control lines are active low. The benefit of having the enable line is that we are able to disconnect a register without doing any physical uncoupling of links etc.

2 Read/write. The terms 'read' and 'write' are used to describe the direction of data movement. We 'write' data into a register then 'read' the data to recover it.

You may remember that the flip-flop in Figure 6.1 included a separate line for reading the data and another for writing. Now, while this was OK with a single flip-flop, a 64-bit register would require 128 lines just to carry the data in and out. By using the tri-state buffer from Chapter 5, we can use each line to read and write data as required. The tri-state buffers are all controlled by the logic level applied to the read/write line. The normal convention applies – the line over the 'write' means that this line is taken low to write data and, of course, high to read data.

It may be interesting to look inside the register to see how the tri-state buffer is used to achieve this two-way traffic on a single wire. Have a look at Figure 6.6. Two tri-state buffers are connected back-to-back. In the first example, logic 1 input will enable the top buffer. The control voltage is inverted to a logic 0, which then disables the

Figure 6.6

Two-way
data flow

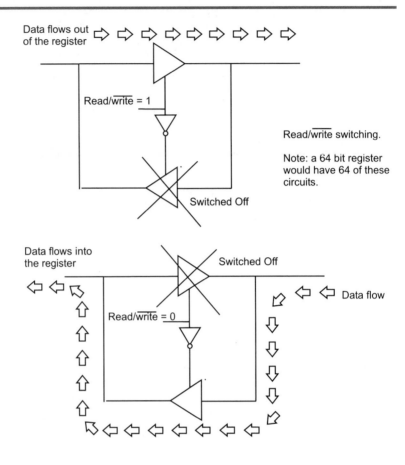

Data flows out
of the register

Read/write = 1

Read/write switching.

Note: a 64 bit register
would have 64 of these
circuits.

Switched Off

Data flows into
the register

Switched Off

Read/write = 0

Data flow

lower buffer. Data can now flow from left to right. When the control
signal changes to a logic 0, the top buffer is disabled and the lower
one is enabled and the reverse direction of data flow is possible.
Note how two buffers and an inverter are used for each line to be
controlled. A single control line is used to switch all the data lines at
the same time.

What are registers for?

Registers are storage areas inside the microprocessor. Almost the
whole of the microprocessor is made of registers. They store the data
that is going to be used, they store the instructions that are to be used
and they store the results obtained. Nearly all registers involve tri-state
buffers to control the direction of data flow.

In most cases, the data to be stored is applied to the inputs of the
register and, after a short pause to let the voltages stabilize, the register
is enabled by the voltage on the enable control. The information is
then safely stored until it is next required.

The sequence is:

1 The read/write line is taken to logic 0 to allow the register to receive data from an external source.
2 The enable control switches ON the tri-state buffers at the input to each flip-flop.
3 The data is written to each flip-flop and then the enable control puts the register to sleep until the next time it is needed.

How long can it be stored?

It will be stored until the power supplies are removed – either by an equipment fault or, more usually, by the system being switched off. The data does not deteriorate in storage.

Shift registers

These are a variation on the register theme. They still consist of group of flip-flops but the interconnections have been changed. Have a look at the arrangement in Figure 6.7 and see if you can guess the likely outcome.

Figure 6.7

A shift-left register

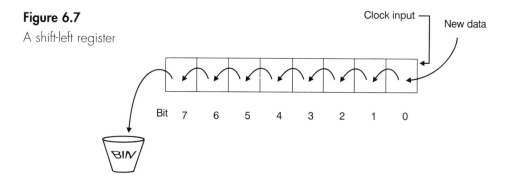

This is called a shift register because the data is shifted from one flip-flop to the next each time the clock pulse occurs. Specifically, the one shown is a shift left register because each bit moves one place to the left on each clock pulse. All the bits move at the same time. The last one in bit 7 drops off the end and is lost while at the other end, a new bit is entered into bit 0.

In Figure 6.8, the register has been loaded with the binary equivalent of 36_{10} or 24H and a series of zeros has been chosen to be loaded at the bit 0 end.

Figure 6.8

A shift-left register in action

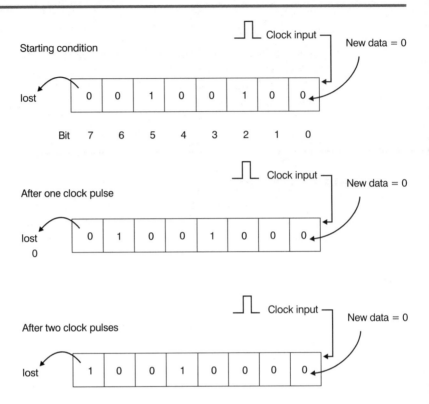

Follow the sequence through and in particular note what happens to the numbers stored:

1 After one clock pulse, all the bits will have moved one place to the left. A new '0' will have entered bit 0 and the last, which was in bit 7, will have fallen off the end of the world. The bits stored at this time are 01001000 and the numerical value is 48H or, in denary, 72_{10}. Notice how shifting the bits to the left has multiplied the value by 2.

2 After eight clock pulses, all the existing data in the register will have been flushed out and refilled with zeros. The register will hold the number zero so there is a limit to how many times we can multiply by shifting the register.

3 After 5000 clock pulses, it is still full of zeros. Admittedly, they will be new zeros that have replaced the others but that will not make any difference.

What happens if we don't apply any input data to enter bit 0?

If the input connection is simply left unconnected, there will be no voltage information coming in to the first flip-flop. The input is said to be 'floating' and will assume some voltage which may be low or high. As the clock pulses are applied this may well result in random data entering the register. Random data is of no help to anyone so we normally overcome this problem by building in a bias in the design of the register to make it have a tendency to move towards one logic level rather than the other. It is up to the manufacturer but most floating inputs will float high and enter ones.

The shift register considered has been a shift-left register, which means that the information is fed in at the right-hand end and moves progressively towards the left along the register until it drops off the end.

By re-arranging the register, it is easy to produce a shift-right register as in Figure 6.9. This has all the same properties except it shifts data towards the right and divides the number by two each time the clock pulse is entered. Compare Figures 6.7 and 6.9.

Figure 6.9

A shift-right register

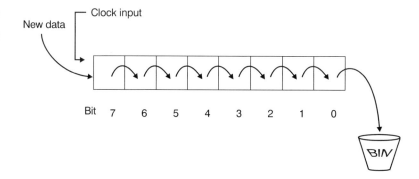

A real world use for a shift register

It is interesting that a shift register can perform simple multiplication and division but it can do many jobs that are more interesting.

One example would be automatic checking of inputs. In Figure 6.10 it is controlling an automatic ticket dispenser. The customer inserts some money and presses any button of the eight available to obtain the ticket required – but which button was pressed?

As a button is pressed the voltage output can be designed to change from logic 0 to logic 1 so to start with, we can assume no buttons are pressed and the response from each button is zero. Along comes a customer who, having read the instructions, inserted some money and re-read the instructions and stared at the buttons, eventually decides to press a button.

Pressing a button generates a burst of eight clock pulses and the value of each button is loaded into the shift register. Once the button has been pressed the zeros and ones corresponding to each of the buttons

69

Figure 6.10

Using a shift register

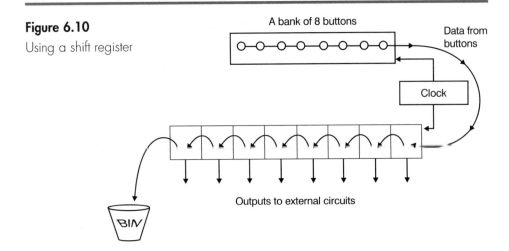

is loaded into the shift register. The output from each button is made available to external circuits and one such circuit will be activated and a ticket will drop down the chute.

For how long would the customer have to press the button?

The microprocessor is amazingly fast compared with us. If we feel the temperature of a piece of metal and it is too hot, we immediately take our hand off. But how long did this take? For most people the time to think and then respond would be about one-tenth of a second. In sport, it means that the person at the receiving end must use body movement or magic to predict what is going to happen. Waiting to respond to the flight of the ball will make them too late.

Most people would therefore press a button for at least 0.1 s. So what can the microprocessor do in the same 0.1 s? A modern micro-processor can check a button in about 0.25 μs or 0.25 millionths of a second. In other words, it can check about 4 000 000 buttons in a second.

We have a best response time of 0.1 s. The microprocessor has a response time of about 0.25 μs. This means than the microprocessor lives at a speed of about 400 000 times faster than us. Can you imagine how we would feel faced with a creature called a 'Waitabit' that moves 400 000 times slower than us? It would take 11 hours to press the button. After all that effort, it may run off at 3 cm/h (1.2 in/h) to spend 11 years having a cup of coffee. By way of compensation, it may well live for 28 million years!

Rotate registers

These are modified versions of the shift registers. There are only two simple changes necessary. The first is that the data is loaded in parallel.

This means that the data is loaded into each flip-flop in the register at the same time. This requires a separate connection to carry each bit but the good news is all the data is loaded under the control of a single clock pulse so it is very much faster. Once loaded, subsequent clock pulses cause the data to be moved along the register as before. The last bit of data is connected back to the other end of the register instead of dropping off the end into our bin. Have a look at Figures 6.11 and 6.12.

Figure 6.11

A rotate-left register

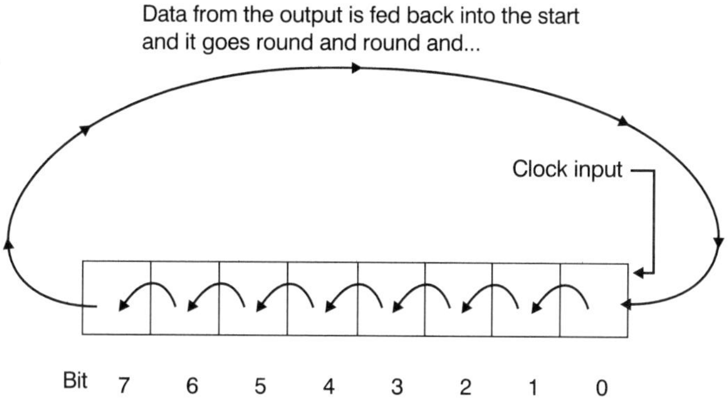

Figure 6.12

Data movement in a rotate-left register

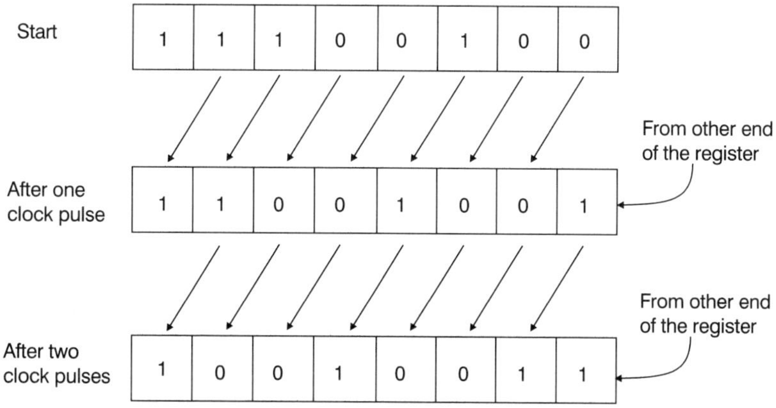

As with shift registers, rotate registers can be made in rotate right as well as left versions. In microprocessors, the same register can be used to rotate or shift in either direction.

The benefit of using a rotate rather than a shift register is that the data is not destroyed. We have seen that a shift register is progressively emptied as bits fall into the bin at the end. With a rotate register, the data is not changed. If we rotate left say, six times, we only have to rotate right six times to recover all the original data.

Memories

The function of a memory is to store information – almost the same as we said for the register. Generally, a register lives within the microprocessor and stores small quantities of data for immediate use and it can do useful little tricks like shift and rotate. A memory is designed for bulk storage of data but that is all it can do – no tricks this time.

Well, almost no tricks – some types can remember the data even when the power is switched off. The ability to remember data after the power is switched off is the dividing line between the two main types of memory. If it loses its data when the power is switched off, then we call the memory RAM or volatile memory. If it can hold on to the data without power, we call it ROM or non-volatile memory (volatile means 'able to evaporate'). This is seen in Figure 6.13.

Figure 6.13

The two classes of memory

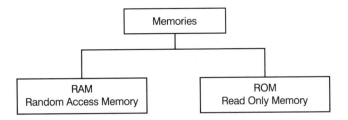

RAM

The letters RAM stands for Random Access Memory which is a silly, out-of-date, name. It should be called read/write memory or RWM but it is so difficult to get something to change once it is established. Anyway, let's leave the name for the moment and look at the memory.

The memory comes in an integrated circuit looking like a small microprocessor and is usually called a memory chip. Inside, there are a large number of registers, hundreds, thousands, millions depending on the size of the memory. Incidentally, when we are referring to memories, we use the word 'cell' instead of register even though they are the same thing.

So, each of the internal cells may have 4, 8, 16, 32, or 64 bits stored in flip-flops. Figure 6.14 shows the register layout in a very small memory containing only 16 cells or locations, each of which can hold 4 bits and is given a memory number or address.

Figure 6.14

The layout of cells in
a memory

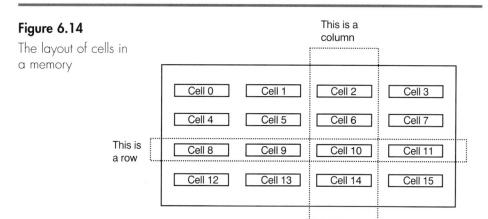

The cells or 'memory locations' are numbered from 0 to 15.
Each holds binary data – anywhere between 1 and 8 bits in each.

This RAM word

In prehistoric computing days, the memory would be loaded in order. The first group of bits would go into location 0, the next would go into location 2, then location 3 and so on rather like a shift register. This meant that the time to load or recover the information would increase as we started to fill the memory and have to move further down the memory. This was called sequential access memory (or serial access memory), abbreviated to SAM. This was OK when a large computer may hold 256 bits of information but would be impossibly slow if we tried this trick with a gigabyte.

To overcome this problem, we developed a way to access any memory location in the same amount of time regardless of where in the memory it happens to be stored. This system was called random access memory or RAM.

All memory, whether volatile or non-volatile is now designed as random access memory so it would be much better to divide the two types of memory into read/write and read only memory. But it won't happen, RAM is too firmly entrenched.

Accessing memory

Each location in a memory is given a number, called an address. In Figure 6.14, the 16 locations of memory would be numbered from 0 to 15, or in binary $0000-1111_2$. The cells are formed into a rectangular layout, in this case a 4×4 square with four columns and four rows.

To use a cell, the row containing the cell must be selected and the column containing the cell must also be activated. The shaded cell in Figure 6.15 has the address 0110 which means that it is in row 01 and in column 10.

To access this cell we need to apply the binary address to the row and column decoders. When the address 0110 is applied, the first half of the address, 01, is applied to the row decoder and the second half of the address is applied to the column decoder. A decoder circuit is a small logic circuit that, when fed with the address of the location, is able to switch on the appropriate row and column.

Figure 6.15

Selecting a memory location

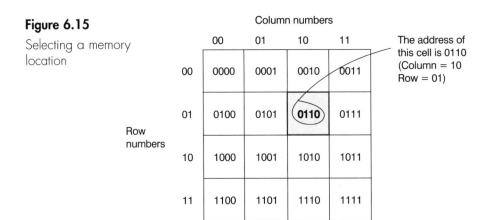

The maximum number of locations that can be addressed will depend on the number of bits in the address. We have already seen that a 4-bit address can access 16 locations. This was because $2^4 = 16$, so, generally 2^n = number of locations where n is the number of bits in the address. To take a more realistic example, if we had 20 address lines we would have $2^{20} = 1\,048\,576$ or 1 Meg locations.

Two types of RAM

Ram chips can be designed in two different forms which we call static RAM (SRAM) and dynamic RAM (DRAM), as seen in Figure 6.16.

Static RAM

These are constructed of flip-flops. The problem with the flip-flop is that it draws current all the time. Therefore, it tends to get rather warm

Figure 6.16

The two types of
RAM

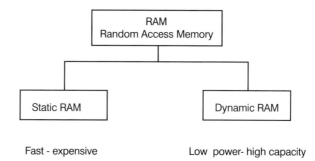

and, on a single chip, the components cannot be packed together very tightly. The benefit is that they are very fast and are used where speed of access is important. Static RAM is often called SRAM.

Dynamic RAM

These store the information in capacitors, which are small components that store an electrical charge in the form of static electricity. They are called 'dynamic' owing to one of its drawbacks. In use, the electricity stored in each capacitor leaks away because of the imperfect insulation. So, after a little while the charge has to be replaced otherwise the DRAM will be empty and all the stored information will be lost. This replacing is called 'refreshing' and has to be performed at intervals of about 2 ms by a DRAM control circuit. To prevent any interference with the operation of the microprocessor system, the refreshing is done in the background whenever the DRAM is not being used.

Once the static charge is stored, no further current is required (except for refreshing), therefore less heat is being generated internally and we can pack more memory into a given space. We say it has a high packing density.

Memory organization

A memory contains a number of cells or registers that, themselves store a number of bits. In Figure 6.14, we saw a really simple memory with 16 locations, each of which could store between 1, 4 or 8 bits. The memory organization is always quoted as 'number of locations x bits stored in each' so this memory would have an organization of anywhere between 16×1, 16×4 or 16×8.

Static RAMs usually store 8 bits in each location so a typical chip size would be $131\,072 \times 8$ giving a total storage capacity of $1\,048\,576$ bits. This is often referred to as $128\,\text{K} \times 8$.

Dynamic RAMs store either 1 or 4 bits in each location. One bit in each is very popular, so a typical chip organization would be $1\,048\,576 \times 1$ which, as we can see, would actually hold the same total number of bits as the example SRAM – it's just the organization that has been changed.

Three types of ROM

All ROMs are used to store information on a more-or-less permanent basis. In use, the ROM can be read but new information cannot be stored in it. In other words, we cannot write to it (see Figure 6.17).

Figure 6.17

Three types of ROM

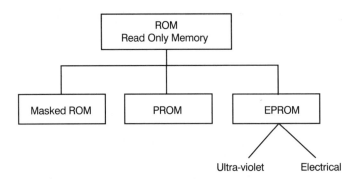

Masked ROM

A masked ROM is manufactured to our specification and cannot be changed. We must be very sure that the information is correct before it is made otherwise it all goes in the waste bin and the person responsible is looking for a new job. The initial cost is necessarily high due to the expense of the tooling required. It is only worthwhile if at least a few thousand identical chips are required (see Figure 6.18).

Figure 6.18

The economics of ROM choice

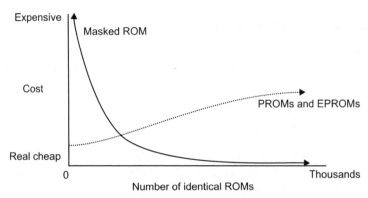

Programmable ROM (PROM)

This chip is supplied with all the data held at zero by means of small internal fuses. When one of the fuses is blown, the associated bit changes from 0 to 1. To blow the fuses a piece of programming equipment is needed. This equipment can be purchased quite cheaply if only one PROM is to be programmed at a time. If a larger throughput is needed then this will inevitably increase the cost of the equipment. Once the fuse is blown, it cannot be repaired so if you make a mistake, the chip is wasted.

The ROM is useful for low volume production because the initial costs are much lower than the masked ROM but you do have to program them yourself.

Erasable programmable ROM (EPROM)

As the name would suggest, this chip allows us to program it, then change our mind and try again. To erase the data there are two methods – ultraviolet light or electrical voltage pulses.

EPROMs are ideal for prototyping since it is so easy to change the data to make modifications.

The UVEPROM

The chip is bombarded with ultraviolet light via a transparent window on the chip. A specially constructed EPROM eraser provides the light. We pop the chip in, close the lid and switch on the timer. After a few minutes, the data is erased. When erased, all the data output is set to 1. We then put the chip into an EPROM programmer, usually the same piece of gear that was used to program the PROM. We can feed in the new data and within a couple of minutes, we have finished the process.

They can be erased and reprogrammed about 700 times before they become increasingly reluctant to erase and their life is over. Once programmed, the data is safe for about seven years. For long term storage, it is best to reload them or, better still, use a masked ROM if available.

A safety note: be extremely careful not to expose your eyes to the ultraviolet light from the eraser. The wavelength of 253.7 nm is very dangerous.

Electrically erasable programmable ROM (EEPROM)

This chip uses electrical voltage pulses as inputs to clear the previous data and is then reprogrammed in the same way as the UVEPROM. It has the added advantage that individual parts of the data can be reprogrammed without deleting everything first as is the case with the ultraviolet version. EEPROM can be found as serial access (SAM), as well as the more usual random access.

The reprogramming can be done while installed in the micro-processor-based system. It does not need a separate programmer. Their disadvantage is that they are slow to program and have a limited number of reprogramming cycles.

Pin layout of an EPROM

Figure 6.19 shows the pin-out diagram for a 1 Mb (1 048 576 bits) EPROM with an organization of 131 072 × 8 bits.

Figure 6.19

Pin out diagram of an EPROM

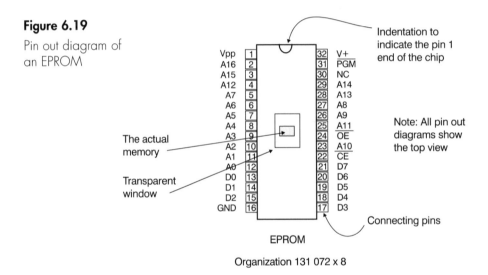

EPROM

Organization 131 072 x 8

Power supplies

The main power supplies to operate the chip are the +5 V applied to the V+ pin and 0 V on the GND (ground) pin.

To program the memory, the programming voltage is applied to Vpp. When not being programmed, it should be held at +5 V. Be careful to read the data book – the value of Vpp differs widely.

Address pins

Address pins are always numbered starting from A0. We have seen that the number of location is given by 2^n so with 17 address lines (A0 to A16) the number of locations would be 2^{17} = 131 072.

Data pins

Like the addresses, these pins always start counting from zero. In the EPROM shown in Figure 6.19, they are abbreviated to D for data and go from D0 to D7 – eight in all. Some manufacturers call them output pins and number them O0, O1, O2 etc. The output from these pins is either 0 V or +5 V or near to these values.

Control pins

1 Chip enable (CE), sometimes called chip select (CS), is the main on/ off switch for the chip. It is usually active low, which means that the chip needs a logic 0 voltage to be applied to switch the chip on. This is indicated by a line over the CE. When the chip is switched off, it goes to sleep and the power drops with a reduction of about 150 times.

2 Output enable (OE) leaves the chip fired up but with its output disconnected from the data pins. This is done by disabling a series of tri-state buffers immediately before the data pins. Disconnecting the output pins is very much faster than switching the chip off. Watch out for the line over the name to indicate the polarities required.

Unconnected pins

These are shown as NC and are not used. They are physically separate from the internal chip and therefore have no effect of anything. They should be left unconnected.

Pin layout of a SRAM

Have a look at the SRAM in Figure 6.20. Many of these pins will be recognized as being the same as we saw with the EPROM.

Figure 6.20

Pin out diagrams of RAM chips

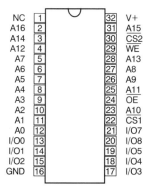

Static RAM
Organization 131 072 x 8 bits

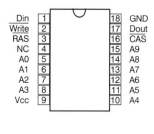

Dynamic RAM
Organization 1 048 576 x 1 bit

Data pins

Since the SRAM is a read/write memory, the data pins are used to read data into the chip and to write data out to the microprocessor system. With this in mind, they have been called input/output pins (I/O) and, as usual, are numbered from zero.

Control pins

1 Write enable ($\overline{\text{WE}}$) controls internal tri-state buffers to control the flow of data to write or read on the data pins. The line over the WE indicates that, to write data, the pin must be held low.
2 There are two chip selects, one shown as active low and one as active high. To enable the chip, both must have the correct voltage present. This provides a degree of flexibility to the system designer. If either is not required, it can simply be held down permanently to its appropriate voltage and then the other can then be used to control the operation of the chip.

Pin layout of a DRAM

Row address strobe (RAS) and Column address strobe (CAS)

At first glance, there does not seem to be enough address pins for the number of locations to be addressed. A0 to A9 is only 10 pins, which would suggest a total of 2^{10} or 1024 locations. The trick here is to use the same pins twice and hence load in a total 2^{20} or 1 048 576 addresses.

The sequence of events dictates that the RAS line is taken low and the bottom half of the address is loaded into the Rows – RAS then returns high. Then the CAS is taken low while the remainder of the address is fed into the columns. After this, the state of the write line (same as R/W) determines whether data is entering or leaving the DRAM.

There is only a single bit stored in each address, the data entering via the data in (Din) pin and leaving via the data out (Dout) pin.

Some more memories that don't fit into the general pattern

SIMMs

Single in-line memory modules are a collection of separate RAM chips that are mounted on a piece of board to make installation quicker and easier. They are not actually a different type of memory.

RAM Cards

The problem with RAM memory is that it is volatile. The information is lost as soon as the system is switched off. RAM cards overcome this by providing the RAM chips with their own on-board battery. In this way, the RAM card can be removed from the system without losing the data. It is really a RAM pretending to be ROM. This provides full speed operation and permanence during the 10-year life of the battery.

Flash memory

This is non-volatile RAM (NVRAM). In fact, it is really a form of RAM with a battery installed to provide power during shutdown periods. Rather like a single chip version of a RAM card.

Memory maps

A microprocessor has a number of address lines that can be used to access RAM or ROM or other devices within the system. As we saw in the memory chips, the total memory addressable by a microprocessor is found by the formula 2^n where n is the number of address lines. For example, an 8-bit microprocessor generally has 16 address lines and can access 2^{16} or would have 65 536 or 64 bytes of memory. The Digital Alpha 21064 has 34-bit address lines giving 2^{34} or a little over 17 Gbytes. This memory is shared between the RAMs, ROMs and other devices, including some for the microprocessor itself to use.

The system designer has to decide in what way the available memory is to be used. Using the memory map of the 8-bit microprocessor as a simple example, we start off with a blank space as in Figure 6.21. When the microprocessor is first supplied with power it will immediately start following the first program provided. How does it

Figure 6.21

The starting point for a memory map

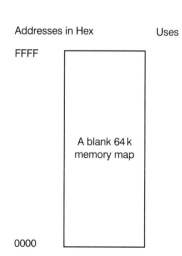

Addresses in Hex

Uses

FFFF

A blank 64 k memory map

0000

Figure 6.22

A typical small system memory map

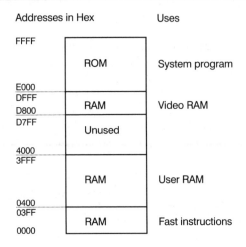

know what program is first? The answer to this is that it goes to a pre-determined address, which has been built into the microprocessor during manufacture.

If we assume the startup address is FFFAH, then we must put some useful information at that address for it to follow. This implies that some ROM memory must be at the top end of the memory map. Most provide some extra fast programming ability and this usually requires that some RAM to be available at the bottom end of the memory map.

There is nothing else that is allocated to any particular address so any other RAM and ROM memories can be placed at any position on the map. The map does not have to be full, indeed it seldom is. The balance between RAM and ROM depends on the purpose to which the system is to be put. A control system running a piece of manufacturing machinery is likely to be predominately ROM whereas a computer would need significant RAM. A simple memory map is shown in Figure 6.22.

Sorting out the addresses

This is just an exercise in hex numbers in which a 'hex' calculator will prove invaluable. Now, 1 kbyte of memory occupies 2^{10} or 1024 locations, which is 400H in hex. The first section of RAM extends from: start address + highest RAM address = 0000H to 03FFH so the highest address in the 'Fast instructions' section is 03FFH. The User RAM extends from 0400H to 3FFFH. How many kilobytes of memory is this?

This is 15 kbytes of memory. This was found by subtracting 0400H from 4000H to give 3C00H and then dividing this result by 400H, the hex equivalent of 1 kbyte. The division is best done by calculator. Remember that the fact that the user-RAM ended at 3FFFH means that the total number of locations, including the first one in 4000H so it was more convenient to use the figure of 4000H straight from the memory map. This is sometimes a little difficult to fully come to terms with, but a cup of coffee and a slump in an armchair often helps.

In a similar way, the video RAM that holds the information to be displayed on a monitor can be found by subtracting the lower address D800H from the higher address E000H to give 800H. Dividing by 400H indicates 2 kbytes of video RAM.

Example

If a 12 kbytes block of ROM started at the address 8000H, what is the highest address in the ROM?

Since 1 kbyte = 1024, it follows that 12 kbytes = 12 × 1024 or 12 288 in denary. Converting this to hex gives 3000H. Now, we have to be a bit careful. If the ROM includes 3000H addresses, they will run from 0 to 2FFFH. Adding the start address of 8000H to the highest address will give 8000 + 2FFF = AFFFH, which is the highest address in the ROM.

Quiz time 6

In each case, choose the best option.

1   An SRAM with 12 address pins and 8 data pins would have:

(a) an organization of 12 × 8 bits.
(b) approximately $16\frac{3}{4}$ M locations.
(c) an organization of 12 × 8 bits.
(d) a storage capacity of 32 768 bits.

2   A bistable:

(a) can store two bits of information.
(b) is another name for flop-flip.
(c) has a floating output.
(d) is made from several registers.

3   The pin that is most similar to one marked as CS may be labelled as:

(a) OE.
(b) CAS.
(c) CE.
(d) Vcc.

4   If, in a memory map, the lowest address of an 8 kbyte RAM is 1000H the highest address would be:

(a) 8192H.
(b) 2FFFH.
(c) 7FFFH.
(d) 3000H.

5   A UVEPROM:

(a) is programmed by ultraviolet light.
(b) loses the data if the power supplies are disconnected.
(c) is a form of non-volatile memory.
(d) is used in a SIMM.

7

A microprocessor-based system

How simple can a microprocessor-based system actually be? It must obviously contain a microprocessor otherwise it is simply another electronic circuit. A microprocessor must be programmed. This means that it must be provided with a series of instructions to be followed. However we program the microprocessor, the result is a series of binary numbers that represent the simple step by step instructions to be followed. These instructions must be stored in some memory. But do the instructions have to be stored in RAM or ROM? It *must* be in ROM. Remember that RAM will hold random data when first switched on and if our microprocessor was controlling the operation of a dynamite factory, the last thing that we would want is for it to start following random instructions at the rate of a million a second!

What determines how fast the microprocessor carries out the instructions? For the moment we will say that regular pulses of voltage applied to the microprocessor determine its speed. This voltage pulse is called a clock pulse.

The clock

A clock circuit controls the operation of the microprocessor. This produces a series of voltage pulses like a ticking clock. The whole system runs sequentially, doing the required jobs one after the other. One step completed for each tick of the clock system.

The clock circuit can be mostly internal to the microprocessor or it can be entirely external. It is unlike a normal watch or clock in that the exact speed is not important. It may be running at a nominal rate of 200 MHz but if it actually ran at 199 MHz or 201 MHz there would be no great panic. With a modern watch, the equivalent error of seven minutes a day would mean a trip back to the supplier. In Figure 7.1 a typical clock pulse is shown.

Figure 7.1

A 200 MHz clock pulse

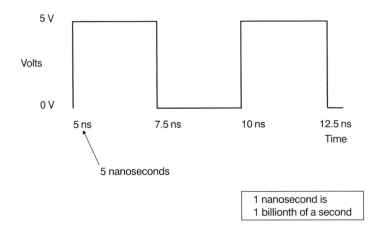

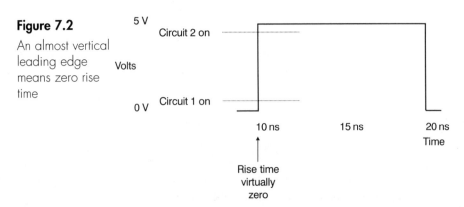

The shape of the clock pulse is stipulated in detail in the microprocessor specification. If it does not stay within the limits stated, serious problems can occur. Let's assume that two circuits have to switch at the same time but they operate at slightly different voltages. In Figure 7.2 the two operating voltages occur at virtually the same moment due to the very fast rise time. In Figure 7.3 the clock pulse has a very slow rise and fall time and Circuit 1 will switch before Circuit 2. In this example, the difference is about 2 ns. Whether or not a 2 ns difference is significant will depend on the circuit being considered.

Figure 7.2

An almost vertical leading edge means zero rise time

Figure 7.3

The rise time is about 2 ns

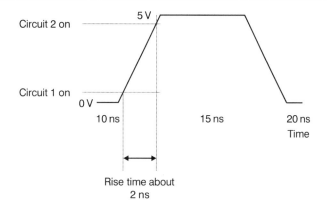

All microprocessor technical data will stipulate the maximum rise time and fall time and between which voltages it is being measured. For example, the Intel 80386 has a supply voltage of 0–5 V and must have a rise time (and fall time) of less than 2 ns when measured between 0.8 V and 4.2 V.

If the rise time (written as t_r or t_{rise}) is not shown, it is normally measured between 10 and 90% values of the supply voltage. Each microprocessor has a maximum frequency for the clock pulses and a minimum value.

Why is there a minimum speed?

Microprocessors have internal registers and storage areas called capacitors that are like dynamic RAM and need refreshing at intervals. As the clock speed is reduced, the interval between refreshing gets longer until the register or capacitor can no longer hang on to the information and the whole operation collapses. Typically, the minimum clock speed is about a quarter of the maximum speed. The Intel 80386 will run down to 800 kHz.

The first microprocessor, in 1971, was the Intel 4004. It ran at a clock speed of 0.108 MHz and handled 4 bits at a time. In 1998, the 64-bit Alpha 21164 runs at 600 MHz, king for a day but will, no doubt, soon take its place on the museum shelf. (It is 700 MHz now and the ink is barely dry!)

The most basic system is therefore going to include a microprocessor, clock and a ROM chip to provide the built-in sequence of instructions. The only other essential is to have wires or other conductors to connect the circuit. Figure 7.4 shows the most basic microprocessor-

Figure 7.4

The most basic microprocessor-based system

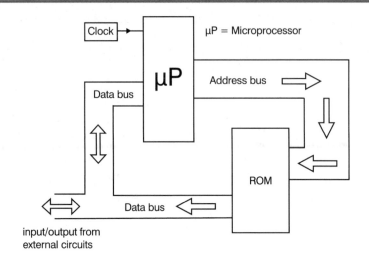

based system with just the bare essentials. As usual, the power supplies are not shown.

All microprocessors are connected internally and to the rest of the system by collections of conductors called buses.

Buses

These conductors are in groups, since many will be going to the same place for much the same purpose. For example, an 8-bit micro-processor normally uses 8 connectors to carry the data between the microprocessor and the memory. It would make diagrams very complicated if each wire were to be shown separately so we group them together and refer to them as the data bus. A bus is therefore a collection of conductors providing a similar function.

In a microprocessor-based system we have three main buses: the data bus, the address bus and a control bus. The data bus is a two-way bus carrying data around the system. Information going into the micro-processor and results coming out. The address bus carries addresses and is a one-way bus from the microprocessor to the memory or other devices. The control bus is rather different from the other two. It is a somewhat looser collection of conductors. If we look at a micro-processor-based system we can easily see the data and address buses since they consist of many parallel connections. However, the control bus is just an association of all the other necessary connections such as those to the chip select and read/write pins.

Address information, data and control signals have to be carried around inside the microprocessor as well as in the external system. We will therefore meet internal as well as external buses.

Input/output circuits

The signal on the data bus has only a very low power level and to be of use it must be amplified.

The first response was to put a series of amplifiers; one on the end of each of the connections on the data bus but this was soon superseded by a more sophisticated chip with more facilities. For example, if we wanted to send a signal to a printer at 1 ms intervals it would not be sensible to tie up the main processor with counting out these time intervals. It would be much better to tell the output chip to do its own timing and thus release the microprocessor for more important jobs.

The output devices have become quite complex and now go under a variety of names:

I/O controller	– input/output controller
PIA	– programmable interface adapter
VIA	– versatile interface adapter
PIO	– programmable input/output
PPI	– programmable peripheral interface

. . . and many others.

Whatever name is given to the device, they are basically the same and are essential to any microprocessor-based system. An improved basic system is shown in Figure 7.5.

Figure 7.5

An input/output chip has been added

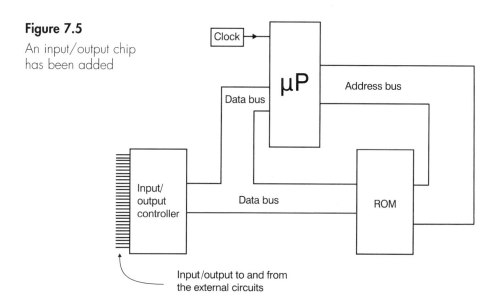

Input/output to and from the external circuits

A practical microprocessor system

Apart from in the most minimal of circuits, some RAM is needed. Even if the microprocessor-based system is controlling an oven, we still need the facility to vary the instructions to change the temperature, the time cycle, the fan speed etc., so some RAM must be added. Some microprocessors have a small amount of RAM included internally, enough for this sort of system to work but still quite limited.

If we add some external RAM, the microprocessor is controlling the operation of three chips: ROM, RAM and I/O. To control the flow of information it needs to send chip select and read/write information along the control bus.

Read/write signals tell the RAM and the I/O chip whether they have to read, i.e. accept information from the data bus or to write information onto the data bus.

Chip select is the on/off switch for each of the chips and we have to be very careful to ensure that only one set of information is being connected to the data bus at one time.

If a ROM chip were to be putting a binary 0 onto one of the data bus connections and, at the same time, another ROM or a RAM chip was applying a binary 1, there would be a disagreement between the two chips. What would happen? It is sad but the likely outcome is a fight to the finish with one or other of the chips being condemned to the waste bin.

To prevent this from happening, an address decoder circuit samples the address bus and selects the appropriate chip. If the microprocessor wishes to send some data to the RAM chip, it applies a suitable address to the address bus that is applied to the ROM, RAM and the I/O controller but there are no problems at the moment since all these chips are switched off. The address decoder applies the inputs from the address bus to an array of logic gates that have been organized to comply with the memory map of the system. The output from the address decoder then switches the RAM chip on and the ROM and I/O chips off. The design of the address decoder can be modified to control any number of external chips in the system. An upgraded system is shown in Figure 7.6.

How it all works

To demonstrate its operation we can ask it to perform a simple task.

Instruction: Send the number 25H which is in the ROM and store it in the RAM at address 2500H. This is what happens – follow the action on Figure 7.6.

1 The microprocessor has to collect the instruction from an address in ROM. It does this by putting the address onto the address bus.

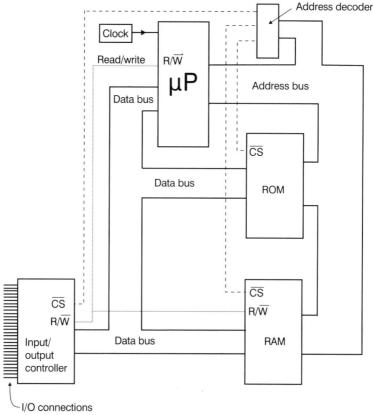

Figure 7.6

A complete microprocessor system

2 The address is applied to the ROM and the RAM as well as the address decoder. This will not cause any problems because all the chip selects will be switched off at the moment. When the logic gates within the address decoder responds to the input from the address bus the result will be that the ROM is switched on and the other two are kept off.

3 Switching on the ROM will mean that it takes in the address from the address bus. Inside the ROM chip, the row and column decoders activate one of the memory locations and the binary number stored at that location is placed on the data bus by switching on the tri-state buffers. As soon as the information is read, the chip select will switch the ROM chip off.

4 The information which is now on the data bus is read by the microprocessor. It is an instruction which can be interpreted as 'go to address F600H and read the number that is stored in that address'.

5 In response to this instruction, the microprocessor puts the address F600H onto the address bus.

6 The address decoder applies this number to its logic gates and this results in the chip select of the ROM chip being switched on again. The ROM chip accepts the address F600H into its row and column decoders and then puts the number 25H onto the data bus.

7 This number is stored temporarily in the microprocessor.

8 The microprocessor then puts the number 2500H onto the address bus and the address decoder puts a signal on the chip select of the RAM chip to switch it on. It then sends a logic 1 on the read/write line. The RAM is switched on and it is told to read the data on the data bus. The read/write line goes to the I/O chip as well but again, this causes no problem because its chip select line is keeping it switched off.

The number 25 is now safely stored in the RAM chip and will remain there until it is over-written with new information or the power is switched off.

Another look at the address decoder

We have seen in Chapter 6 that the number of locations that can be addressed is 2^n where n is the number of address lines. By feeding the numbers into our calculators we can see that the relationship between lines and locations is as shown in Table 7.1.

We can also use this table to identify the number of lines needed to access a known number of address locations. For example, if we wanted to access 700 locations, we can see that 9 lines could access 512 locations which is too few. Therefore, we would have to go to 10

Table 7.1 Larger memories need more address lines

Number of address lines	Number of locations	Number of address lines	Number of locations	Number of address lines	Number of locations
1	2	13	8k	25	32M
2	4	14	16k	26	64M
3	8	15	32k	27	128M
4	16	16	64k	28	256M
5	32	17	128k	29	512M
6	64	18	256k	30	1024M=1G
7	128	19	512k	31	2G
8	256	20	1024k=1M	32	4G
9	512	21	2M	33	8G
10	1024 = 1k	22	4M	34	16G
11	2k	23	8M	35	32G
12	4k	24	16M	36	64G

lines which would actually give access to 1024 locations. The 'real' answer of 9.45 is not sensible because we cannot have 0.45 of a connecting wire so if 9 is not enough, it will have to be 10.

For those who like to see the calculations, the mathematical result is given by:

$$\text{number of address lines} = \frac{\log_{10}(\text{number of locations})}{\log_{10}(2)}$$

Designing a decoding circuit

Let's imagine that we have a microprocessor-based system using eight memory chips, ROM or RAM it doesn't matter. Each of the chips holds 8 kbytes of memory. From Table 7.1 we can see that an 8 kbytes chip will require 13 address lines in order to access each of their internal locations. Assume too, that the microprocessor that we are using has a 16-bit address bus so we have the situation shown in Figure 7.7. The address lines are numbered from A0 (address line number 0) to A15. The 13 bits A0–A12 are heading off towards the ROM and RAM chips. The remaining three address lines, A13–A15, are used by the address decoder.

Figure 7.7

There are three 'spare' address lines

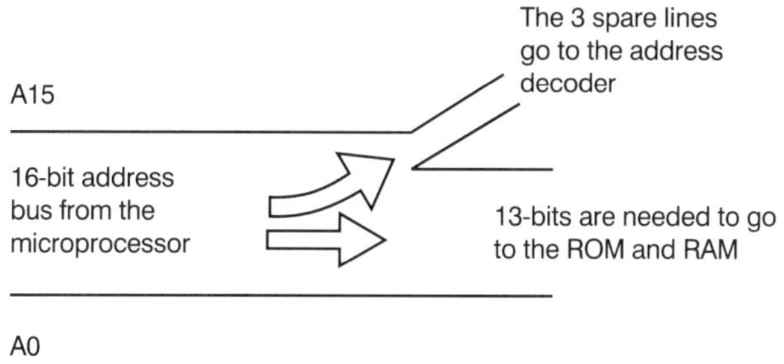

The decoding chip

The decoder circuit can be made from separate logic gates or can be bought ready-built in a single integrated circuit. For ease of construction, most designers opt for this choice for the result is smaller, dissipates less heat and is less expensive (and it works first time). There is very little to be said for the build-it-yourself approach.

The basic requirements are three input address lines and eight output lines each connected to one of the chip select pins on a memory chip.

To switch the chips on, the chip select must be taken to a logic 0 voltage. A logic 1 voltage level will switch the chip off. It is vital, of course, that only one chip can be switched on at the same time otherwise they will load competing data onto the data bus and are likely to be destroyed. The three addresses can result in $2^3 = 8$ different inputs to the logic gates

93

Figure 7.8

The operation of
the address
decoder

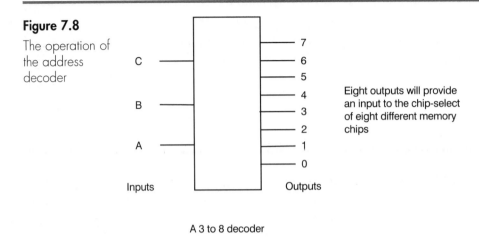

Eight outputs will provide
an input to the chip-select
of eight different memory
chips

A 3 to 8 decoder

built into the decoder chip. The internal design ensures that when the
address pins are all at zero, the first output goes to a logic 0 and all the
others remain high. The memory chip to which this first output is
connected is switched on and all the others are on. When the next
combination of inputs 0, 0, 1 is applied, the second memory chip is
switched on and the others are off. The next combination switches on
the next memory chip and so on until the three input wires have
switched on each of the memory chips with a single combination of
addresses (see Figure 7.8). With three inputs and eight outputs, it is
referred to, reasonably enough, as a 3 to 8 decoder.

Table 7.2 looks a lot worse than it really is. It is really just a
summary of the decoder chip outputs. If the microprocessor put
the address C2F1H on the address bus, then in binary it would be:
1100 0010 1111 0001. It has been broken up into groups of four

Table 7.2 The 3–8 decoder can control eight memory chips

Inputs			Outputs								Results
C	B	A	0	1	2	3	4	5	6	7	
0	0	0	**0**	1	1	1	1	1	1	1	Chip 0 selected
0	0	1	1	**0**	1	1	1	1	1	1	Chip 1 selected
0	1	0	1	1	**0**	1	1	1	1	1	Chip 2 selected
0	1	1	1	1	1	**0**	1	1	1	1	Chip 3 selected
1	0	0	1	1	1	1	**0**	1	1	1	Chip 4 selected
1	0	1	1	1	1	1	1	**0**	1	1	Chip 5 selected
1	1	0	1	1	1	1	1	1	**0**	1	Chip 6 selected
1	1	1	1	1	1	1	1	1	1	**0**	Chip 7 selected

Table 7.3 In full decoding, every address line is used

A15	A14	A13	A12	A11	A10	A9	A8	A7	A6	A5	A4	A3	A2	A1	A0
1	1	0	0	0	0	1	0	1	1	1	1	0	0	0	1

This selects chip 6

This address is internally decoded within the memory chip and points to a single memory location

just to make it a little easier to read. The most significant bit, A15, is on the left-hand end.

In Table 7.3, we can see that 13 out of the 16 address lines go to the memory chip and the other three are fed to the decoder chip. The three lines going to the decoder chip carry the data 1 1 0. We can see that the values C = 1, B = 1 and A = 0 occur near the bottom of the table. These values result in Chip 6 receiving a logic 0 value and thus being selected for use. All other chips are deselected by the logic 1.

Full and partial decoding

Full decoding

In the above example, the 8 kbyte memory chips used 13 address lines and the decoder used three. This makes a total of 16 lines used out of a 16-bit address bus. There are no unused lines and this is referred to as 'full decoding'.

Partial decoding

Now let's make a small change. The memory chips used are 4 kbyte each rather than 8 kbyte. What effect would this have?

The first result would be that the number of address lines going to the chips would be reduced to 12. There are still only eight chips to be selected so a 3–8 decoder is enough. So what have we got now? Twelve address lines to the chips and three to the decoder and one left over and unused. If nothing is connected to this line, then it cannot matter what voltage it carries (see Table 7.4).

We will look at our previous address C2F1H. If it happened to go to a value of 1, the address would change from:

1 1 0 0 0 0 1 0 1 1 1 1 0 0 0 1 (C2F1H)

to

1 1 0 [1] 0 0 1 0 1 1 1 1 0 0 0 1 (D2F1H).

95

Table 7.4 One address line is unused

A 15	A 14	A 13	A 12	A 11	A 10	A 9	A 8	A 7	A 6	A 5	A 4	A 3	A 2	A 1	A 0
1	1	0	X	0	0	1	0	1	1	1	1	0	0	0	1

←——————→ 3 lines for decoding

←——————————————————————————→ 12 lines to the memory

This line is unused so it can have a value of 0 or 1

We now have two numbers that can be placed on the address bus which will result in access to the same memory location since all the bits that are actually used are identical. If we instructed the microprocessor to store some information in the address C2F1H and then to recover the information from address D2F1H, we would get the same information again. The address D2F1H is referred to as a ghost address or an image address. It is important to appreciate that ghost or image addresses have no effect at all on the operation of the microprocessor system. They are merely alternative names for a single address. Incomplete or partial decoding always gives rise to image addresses, their number and their addresses are easily worked out. In technospeak, we say that partial decoding results in more than one software addresses pointing to the same hardware address.

A worked example

The eight memory chips were now changed to 1 kbyte chips. Some data is stored in the address 4000H. Find (a) how many image addresses will occur and (b) the image addresses.

(a) A 1 kbyte memory chip would have 10 address lines. This was taken from Table 7.1 and the decoder chip still needs three inputs. The 16-bit address we were given, 4000H would result in three unused lines as in Table 7.5. These three binary digits can take on the values 000, 001, 010, 011, 100, 101, 110 and 111 which results in eight different values. This could have been done the slick way by saying $2^3 = 8$. Putting these eight numbers into the address will result in eight different addresses.

One real address and seven ghosts or images. But which is the ghost and which is the real one? It doesn't matter. We can assume

Table 7.5 There are now three unused address lines

A 15	A 14	A 13	A 12	A 11	A 10	A 9	A 8	A 7	A 6	A 5	A 4	A 3	A 2	A 1	A 0
0	1	0	1	1	0	0	0	1	0	1	0	0	1	0	1

To the decoder

Three unused lines

To the 1 kbye memory chip

anything we like since they all point to the same physical memory. Having said this, most people seem to opt for the lowest address as the 'real' one.

(b) Now, to find the actual ghost addresses, all we have to do is to feed in all the binary options, 000, 001 etc. into the 'real' address to generate each of the ghost addresses. This is shown in Table 7.6.

The resulting addresses are 4000H, 4400H, 4800H, 4C00H, 5000H, 5400H, 5800H and 5C00H. Strictly speaking, we should, perhaps, only give the last seven addresses since the question asked for the image addresses and the first one is actually the real address. Notice how the addresses increase in a definite pattern. It always works out this way.

Image addresses and the memory map

You may remember the memory maps that we looked at in the last chapter. They showed the addresses taken by the various memory

Table 7.6 Generating the image addresses

A 15	A 14	A 13	A 12	A 11	A 10	A 9	A 8	A 7	A 6	A 5	A 4	A 3	A 2	A 1	A 0	Address
0	1	0	0	0	0	0	0	0	0	0	0	0	0	0	0	4000
0	1	0	0	0	1	0	0	0	0	0	0	0	0	0	0	4400
0	1	0	0	1	0	0	0	0	0	0	0	0	0	0	0	4800
0	1	0	0	1	1	0	0	0	0	0	0	0	0	0	0	4C00
0	1	0	1	0	0	0	0	0	0	0	0	0	0	0	0	5000
0	1	0	1	0	1	0	0	0	0	0	0	0	0	0	0	5400
0	1	0	1	1	0	0	0	0	0	0	0	0	0	0	0	5800
0	1	0	1	1	1	0	0	0	0	0	0	0	0	0	0	5C00

devices and the unused addresses in between them. When we have partial decoding, it gives rise to image addresses. Now, if we put these image addresses on the memory map a curious thing occurs. Every image address always falls in one of the unused areas on the map so they will never cause problems by clashing with an existing memory chip.

Quiz time 7

In each case, choose the best option.

1 The Digital Alpha 21064 microprocessor has a 34-bit address bus which can access a memory of:

(a) 64 Gbytes.
(b) 64 Mbytes.
(c) 16 Gbytes.
(d) 160 Mbytes.

2 The data bus:

(a) is bi-directional.
(b) consists of eight conductors.
(c) can be used to carry data to RAM and ROM memory chips.
(d) is connected to the chip select pins.

3 A 16-bit address bus is carrying the address 4567H and is partially decoded with lines A15 and A13 being unused. Which one of these addresses would access a different hardware location:

(a) E567H.
(b) 4567H.
(c) 6567H.
(d) 8567H.

4 Which one of the following is NOT essential in a microprocessor-based system:

(a) an address bus.
(b) ROM.
(c) a clock signal.
(d) RAM.

5 Image addresses are:

(a) also called ghost addresses.
(b) due to several hardware addresses pointing to the same software address.
(c) the same as partial addresses.
(d) caused by full decoding.

8

A typical 8-bit microprocessor

The 8-bit microprocessor is described in some detail to provide a foundation upon which others can be built since the story of microprocessors is one of evolution rather than 'bolts out of the blue'.

An 8-bit microprocessor – the Z80180

This microprocessor is the modern version of an old favorite called the Z80 which was developed in 1978 which was itself just an improved version of the Intel 8080. There are many similarities since some of the engineers that designed the 8080 left Intel and went to work for Zilog to build the Z80 which proved to be one of the most popular and longest living microprocessors – it is still widely used even in mobile phones.

When Zilog developed the Z80180, they were naturally anxious to hold on to their previous customers by making the new version compatible with the earlier Z80. This allows customers an easy and inexpensive way of updating their system without the high design costs of starting afresh with a completely different device.

The Z80180 can run any programs that were written for the Z80 as well as adding some new features. Technically, this is referred to as providing full backwards compatibility.

The X in Z8X180

If we were to go to a Ford dealer to buy a truck, they will offer us a whole series of them, all of which are basically the same but with a choice of engine sizes, gearboxes, seating arrangements and many other options. In this way, they can please the maximum number of customers without incurring significant redesign costs.

Microprocessors are much the same. They start with a basic design and similar versions are indicated by a slight change in the type number. So far, Zilog have produced the Z80180, Z8S180, and the Z8L180. By comparing the type numbers, we can see that they only differ in the third number or letter and by replacing this letter by X we can refer to a generalized type as the Z8X180. This use of X is in general use throughout the industry. The group of microprocessors are referred to as a family.

Despite the push from the computer industry for faster and faster microprocessors, there are many purposes for which small, slow and simple microprocessors is ideal. Billions of 4- and 8-bit microprocessors and microcontrollers are whirring away embedded inside things like microwave ovens, printers and instruments. Most of us share our homes with about fifty of these devices without being aware of their existence. They massively outnumber the well publicized high-speed computer microprocessors.

Inside of the central processing unit (CPU)

Figure 8.1 shows the main parts of the Z80180 CPU. It consists of a number of different registers which store or make use of the data when

Figure 8.1

The central processing unit (CPU)

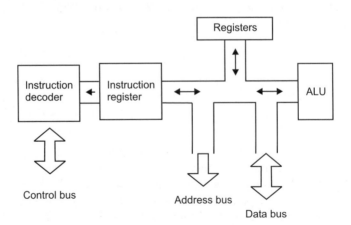

Control bus

Address bus

Data bus

carrying out the instructions in the program. In order to maintain backward compatibility, most of the registers use the same system of labelling with letters to keep it familiar to anyone who has used the Z80.

Instruction register

The instruction register is just a small memory able to store 8 bits, or 1 byte of information. This information is an instruction for the microprocessor to carry out. The information is latched into the instruction register to release the internal data bus for other purposes. It doesn't do anything with the binary input – it just remembers what it is.

Instruction decoder

The instruction decoder is the part of the microprocessor that is able to actually carry out an instruction. Its first step is to identify which instruction that has been entered. It does this by comparing its binary code with an internally stored list. Once it has located the instruction it follows a built-in program called a microprogram. This microprogram is designed into the microprocessor by the manufacturer and details all the necessary steps to complete any instruction of which it is capable. This is the commercially sensitive and critical part of the microprocessor.

When the Z80180 is given a job to do, it must be given an instruction. An instruction is in the form of an 8-bit binary number. We will be coming back to this in a little while.

Arithmetic and logic unit (ALU)

This is a simple pocket calculator. It can add, subtract, multiply, divide, and do various tricks with logic gates AND, OR, XOR etc. Compared with the average scientific calculator, it is pretty poor.

CPU register banks

Technical information about a microprocessor will always include a diagram or listing of the internal registers. Registers are small memories, of between 8 and 128 bits each, depending on the microprocessor being considered. In the Z80180, they are all 8 or 16 bits in size.

Each microprocessor has its own collection of registers so it is essential to read the technical information carefully to see what they do.

There are 'general purpose registers' that are under the control of the user and can be used for a variety of temporary storage purposes.

There are also special purpose registers that are dedicated to particular functions, like the instruction register, for instance.

The Z80180 has sixteen general-purpose registers, and six for special purposes. Figure 8.2 shows the registers.

Figure 8.2

Z80X80 internal registers

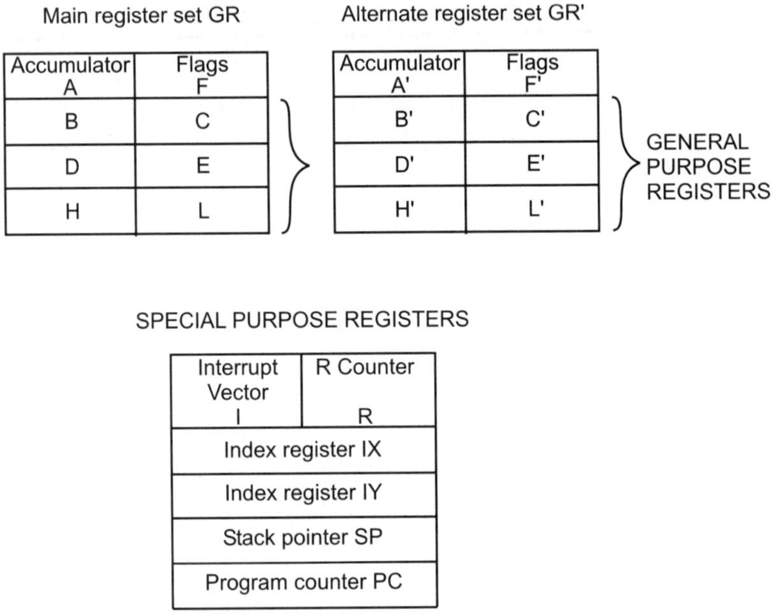

Main register set GR

Accumulator A	Flags F
B	C
D	E
H	L

Alternate register set GR'

Accumulator A'	Flags F'
B'	C'
D'	E'
H'	L'

GENERAL PURPOSE REGISTERS

SPECIAL PURPOSE REGISTERS

Interrupt Vector I	R Counter R
Index register IX	
Index register IY	
Stack pointer SP	
Program counter PC	

We will concentrate on the upper two registers for the moment. The Main register set and the alternate register set are identical except for the small ' against the letters. This ' is called 'prime' so D' would be read as 'dee prime'. The prime is only there to distinguish between the two sets of registers.

Why have two sets? Having two sets allows the programmer to use the main set of registers while the prime registers can be loaded with other information ready for immediate use when required.

Imagine the microprocessor is busily controlling a printing machine when the fire alarm is activated. When the fire alarm signal reaches the microprocessor, it immediately abandons the printing program and starts to run the 'Fire Alarm' program alerting the fire fighters, evaluating the building etc. Some of the information for running this program could be stored in the spare set of registers. When the alarm program has been completed, the microprocessor calmly switches back to its main register set, and continues its printing work until it melts in the fire.

The accumulator

The programmer can use it at any time to store an 8-bit binary number. It can also store numbers to be used in arithmetic operations or for logic operations like AND, OR etc. The results of such calculations are put back into the accumulator after completion. Being only an 8-bit register, it can only hold one byte at a time so any previous data stored in this register will be overwritten as soon as anything else is stored.

The flag register

This is an 8-bit register but it is unusual in that each bit stored is quite independent of all the others. In all other registers, each bit is just part of a single binary value so each bit has a numerical value.

A status register or flag register is like a window for us to see into the workings of the microprocessor. It's rather like a simple communication system. For example, it allows us to say to the microprocessor, 'if you see a negative number during your calculation, just let me know'.

We do not have to take any notice of the flags, they are just indicators that we may take note of or ignore, just as we wish – rather like a thermometer on the wall.

Each flag is identified by a letter and these are shown in Figure 8.3. You will see that two of the bits, 3 and 5, are marked with an X to show that they are not used. These bits were left spare for later development (which has never happened).

Figure 8.3

The Z80180 flag register

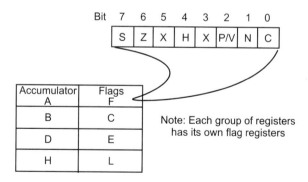

A note: In microprocessors, computers and electronics generally, something that is marked with an X is not used, and so we don't care what the voltage or binary value is.

Another note: a flag is said to be 'set' when its value is 1 and is 'cleared' to 0.

Sign flag (S)

The S flag is just a copy of the bit 7 of the accumulator. A negative number has a 1 in bit 7 and a positive number has a 0 in bit 7 so this flag indicates the sign of the number. You may remember that signed magnitude numbers use a 1 to indicate a negative number and 0 to indicate a positive number. Likewise a negative number expressed in two's complement form will have its left-hand bit as a 1. The number zero is treated as a positive number so we cannot use the S flag to spot a zero result. We do this by employing a special 'zero' flag as we will see in a moment.

Zero flag (Z)

The Z flag spends all its time watching for a result of zero.

You may remember that when we were looking at the uses of an XOR gate, we used this gate to check the code number in a cash dispenser. When the entered value and the card number were the same, it allowed cash to be withdrawn. What actually happened was that the two numbers were compared and if they were the same, then the result would be zero and the Z flag would spring into action. It 'sets' or changes to a binary 1 if it sees a zero result and stays at binary 0 at all other times.

Two things to be careful about. The Z flag only goes to a one state if all bits in the latest result are zero. Also, be careful about the way round: result 0 makes the Z flag = 1; result not zero, Z flag = 0.

All these results would give a Z flag value of zero:

```
0100 0000
1111 1111
0000 0001
1001 1010
```

Only a result of 0000 0000 would make the Z flag go to a binary 1.

Add/subtract flag (N)

The N flag is just there to tell us whether the last arithmetic instruction was 'add' or 'subtract'. N = 0 for add, N = 1 for subtract. Not very exciting.

Carry flag (C) and the half-carry flag (H)

When addition is carried out, it sometimes results in a bit being carried over to the next column. The C flag copies the value of the carry from bit 7 and the H flag records the carry from bit 3 into bit 4 during additions. It also reflects the value of the 'borrow' in subtractions.

Here is an example of an addition showing the C and H flags.

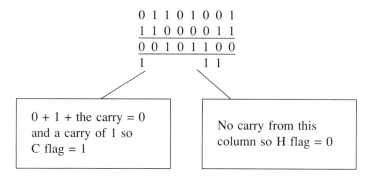

```
0 1 1 0 1 0 0 1
1 1 0 0 0 0 1 1
0 0 1 0 1 1 0 0
1             1 1
```

0 + 1 + the carry = 0
and a carry of 1 so
C flag = 1

No carry from this
column so H flag = 0

Parity/overflow flag (P/V)

This is used for two separate purposes.

The 'V' in the title refers to oVerflow in arithmetic calculations. Some combinations of numbers can be misinterpreted and give an apparently incorrect answer, as we will see.

When the flag spots a possible error it is set, when no error is likely, it is cleared.

Here is an example where we could see a problem.

We are going to add two positive numbers +96 and +66 so even if we can't guess the result, we do at least know that it is a positive number.

```
0 1 1 0 0 0 0 0 = 96
0 1 0 0 0 0 1 0 = 66 +
1 0 1 0 0 0 1 0 = –34??
1
```

The sign bit indicates a
negative number

In this example, we have added two numbers that start with zero and are therefore positive and the result starts with a one, which is a negative number so the P/V flag will be set.

An easy way to check for a possible error is to look at the carry into bit 7 and the carry out of bit 7.

The rule is:

Carry in and carry out
or = No error so P/V flag = 0
No carry in and no carry out

In our example, there was a 1 carried into bit 7 from the previous column but there was no bit carried out from bit 7. This does not fit into the rule above so the P/V flag will be set to 1 indicating a possible error in the result.

Its second purpose is to operate in its parity mode.

In this mode, it checks a byte of data and counts the number of 1 states, if the total is an even number, the P/V flag is set and if odd, it is cleared.

So the byte 10110101 would mean P/V = 0 since the number 1 occurs 5 times, which is an odd number. And 10010110 would result in P/V = 1 because there are four occasions when the number 1 occurs and 4 is an even number.

This flag operates in the P or V mode depending on the instruction being carried out at the time.

We will look at parity in more detail in Chapter 17.

The general-purpose registers

The general-purpose registers are all 8-bit registers but if we like, we can use them two at a time as 16-bit registers. When pairing them up, our choice is restricted to the way suggested by Figure 8.2. We can combine BC, DE or HL but it's not a free choice, we cannot choose any two such as B and L. We can use a pair like BC as a 16-bit register and, at the same time, use D/E/H/L as separate 8-bit registers. As usual, the alternate register set behaves in exactly the same manner.

They have labelled the first four as B, C, D and E, so it may seem a little odd to use H and L for the last two.

The reason for this is that these registers are also used to keep track of memory addresses. The Z8018X has a 16-bit address bus and so requires a 16-bit register to be able to store a full address. The H and L stand for High and Low bytes of the address so if we wanted to store the address 2684H, we would put 84 in the L register and 26 in the H register, using binary of course. As well as being used to hold an address, the H and L registers can still be used for any other purpose that we want, just like B,C, D and E.

Special-purpose registers

Program counter (PC)

A program is a list of instructions for the microprocessor to carry out. Before use, they must be stored in some ROM or RAM. Let's assume that we used addresses starting from 6400H. To run the program we must tell the microprocessor to 'go' from 6400H.

What we are actually doing is loading the address 6400H into the program counter and the microprocessor starts from there. Once it has completed the instruction in 6400H, it goes to the next address 6401H and then 6402H etc until it reaches the end of the program.

The purpose of the program counter is to keep track of the address that is going to be used next.

Stack pointer (SP)

If someone gave us a telephone number to remember, we would be likely to reach for a scrap of paper and a pencil to write it down. Our note may look like Figure 8.4.

Figure 8.4

A number to remember

If we were then given another couple of numbers, we are likely to jot them down, in order, under the first one as in Figure 8.5.

Figure 8.5

A few more numbers

We could read them back in order by reading the bottom one first, then the next and finally the top one. The first one to be entered was the last one to come out.

The microprocessor could do something similar by storing information in a series of 16-bit memory locations called a 'stack'. The stack is loaded in order and then unloaded starting from the last number stored and working back to the top of the stack. The first number that was stored would be retrieved last and for this reason it would be called a 'first-in, last-out' method or FILO. For maximum confusion, this can also be called a LIFO system that, of course, means exactly the same. Have a think about it.

Inside the microprocessor, a series of RAM locations are reserved to be used as a stack and an address counter must be employed to keep track of what address of the stack is to be used next. This counter is called a stack pointer since it 'points at' the next address to be used.

The stack is generally used by the microprocessor to store data on a very temporary basis. We are allowed to use it but the microprocessor takes priority and doesn't know that we have been fiddling with it. For example, if the microprocessor had stored three bits of data in order as in Figure 8.6(a) and we put one piece of our data on the end as in Figure 8.6(b), there would now be data in four locations. However, the microprocessor would not be aware of this. When the microprocessor wished to retrieve its data, it would unload three bytes and accidentally get the wrong one as in Figure 8.6(c).

Figure 8.6

Using the stack

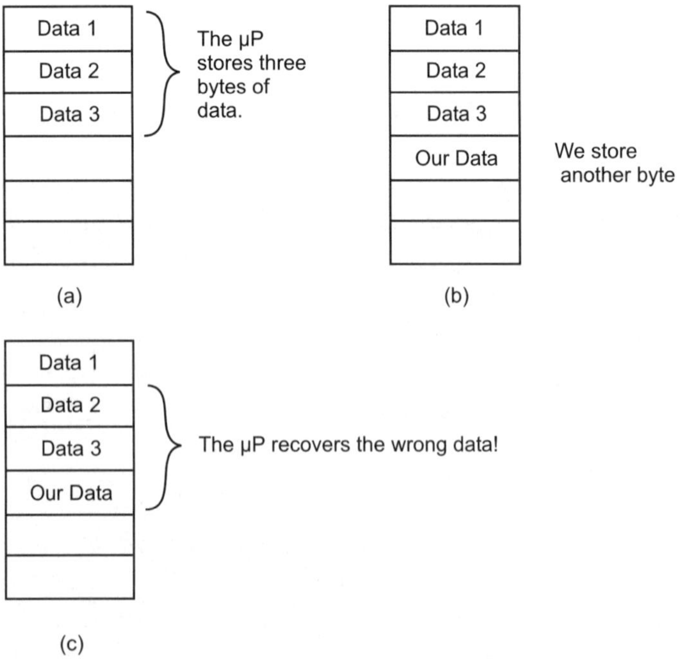

(a) The μP stores three bytes of data.

(b) We store another byte

(c) The μP recovers the wrong data!

The moral of this story is leave the stack alone – or be very careful! The technical data will always give you sufficient information to be able to know whether or not the microprocessor will be using the stack but it still takes a lot of working through to be sure.

Index registers

We have two index registers, one called X and the other called Y. They both perform the same function.

An index is used in situations in which we wish to perform a sequence of similar tasks one after the other. Perhaps to use data stored in a series of memory locations. The 16-bit content of the index register can be added to the contents of the program counter (PC) to produce a new address.

As an example, let's assume that we are running a program which after using address 2600H we would like it to use some data stored in address 3800H. We could do this by loading the 16-bit number 1200H into the X register and instruct the microprocessor to go to the address written as PC + IX (program counter and index X). In this case, the microprocessor goes to address 2600H + 1200H = 3800H to retrieve the data to be used. This jump from 2600H to 3800H is called an offset or we may say that the index X register has provided an offset of 1200H. In this example, index register Y could have been used equally well.

Index registers are often used to input data previously entered into a table or from another part of the memory. The user of a program may be asked to type in a number. This number would be entered in RAM but the program that is going to use this data is stored in ROM. A suitable offset in one of the index registers could shift the address sufficiently to bring it into the RAM area of the memory map.

These registers are also used in situations where the program seeks a response from the user. The screen may show:

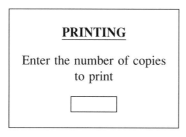

We could implement this by storing the response in one of the index registers and, as the printer performs each copy, the number stored is decreased by one (decremented) until it reaches zero. Every time that the counter is decremented, we instruct the microprocessor to check

the state of the Z flag. When the count finally reaches zero, the Z flag
will be set and the printer can be stopped.

R counter (R)

This is a simple counter in which the lower seven bits are used to
count the number of instructions that have been carried out by the
microprocessor while it has been running the present program. It is
reset to zero whenever the micro is restarted.

Figure 8.7 shows the remaining parts of a microprocessor. All these
parts are inside the single block marked 'microprocessor' (μP) in Figure
7.4. These include all the facilities for connecting the device to
external circuits and a few other essentials.

Figure 8.7

The Z80180
microprocessor

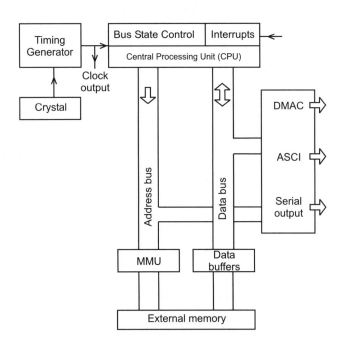

Data buffers

The data buffer is an 8-bit register to store the information being
provided by the external data bus. Once the buffer stores the binary
information then tri-state buffers can lock-in the information disconnecting the external data bus so it can be used for carrying other
data. This process is called 'latching'. As a general rule, all inputs to a
chip from a bus are latched in. This is because the buses take time to
set up new voltages on their lines and need to get started on this job
as soon as possible.

The address and data buses

As with all microprocessors, there are three buses, the address, data and control. As we saw in the previous chapter, the control bus is a loose collection of connections that send signals around the system to make connections and control the operation of each area as necessary. Here we will concentrate on the address and data buses in the Z80180.

The Z80180 has a (sort of)16-bit address bus and an 8-bit data bus.

In the smaller microprocessors, it was usual practice to make the address bus twice the width of the data bus but there is no reason why this has to be the case. The choice of address bus width will depend on the anticipated size of the memory that needs to be addressed in the final designs. Unusually, in this microprocessor there are two conflicting requirements: it has to remain compatible with the 16-bit address bus of the Z80 but we would like to be able to address 1 MB which requires 20 address lines.

These requirements are met by the block shown as MMU or memory management unit.

Memory management unit (MMU)

This circuit includes an 8-bit register that holds address information that can be controlled by the software being used. These 8 bits are combined with the top four address lines, A12–A15, of the address bus. The lower half of these 8 bits are added to the top 4 bits of the address bus and the upper 4 bits are used to expand the useable memory by adding the four extra address lines, A16–A19, to allow a total of twenty lines to be addressed – see note below. The process is shown in Figure 8.8.

The Z80180 family is manufactured in three different packages. The original 40 pins of the old Z80 has increased to 64 pins in the dual-in-line (d.i.l.) package similar to that shown at the top of Figure 1.6. They also make a 68 pin plastic leaded chip carrier (PLCC) version which is just like the square package in Figure 1.6. Finally, there is an 80 pin surface mount type.

Figure 8.8

Expanding the address bus

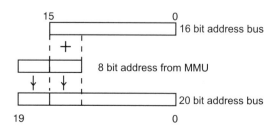

111

Surface mount devices do not have pins that go through holes in the printed circuit board but have connectors that come out horizontally and then soldered onto the surface of the board. This method allows significant space savings.

Note

The smaller number of pins on the d.i.l. layout has limited the number of address pins to nineteen and therefore the maximum memory that can be accessed is now 2^{19} or 512k (really 524288 since a 'k' in this case is taken as 1024 as we met in Chapter 2).

Address buffers

These are tri-state buffers just like the data buffers except they are one-way devices. The microprocessor sends addresses out along the address bus but no address information can come in this way. If we need to send an address into the microprocessor then we are loading information and all information whatever it is goes into the data bus which, as we have seen, can send any data out to the external circuits or accept any data into the microprocessor.

Clock generator

As we saw in Chapter 7, a microprocessor needs a square wave signal to keep all the internal operations in step otherwise it will all end in chaos with data and instructions moving at the wrong times and getting jumbled up.

In this particular microprocessor the clock speed can be set to values between 6 MHz and 33 MHz. The clock signal originates from either a crystal or an external signal source – but never at the same time. Using an external signal can be useful where the microprocessor is used as part of a larger installation and this would allow the microprocessor timing to be governed by the surrounding circuitry.

There are three nice things about increasing the frequency of a crystal: they get smaller, lighter and cheaper. The clock circuitry includes a divide-by-two circuit to allow the use of a double-frequency crystal with the above-mentioned benefits.

Interrupts

You may remember our sad little story of our microprocessor-based system controlling the printer while all around the office was burning. It stopped its main program and went off to phone the fire services, alert the maintenance staff, activate sprinklers etc.

Fire is not the only hazard our little friend can safeguard us from. It could warn for other things like the paper running out in the printer or data corruption on the telephone cable.

Now, we would not want it to send for more paper to deal with a possible problem with the telephone cable so the microprocessor needs to be told what the problem is and what to do about it.

The different programs for each of these problems are stored in a group of addresses in a ROM chip. As an example, we could load the programs at these addresses:

0800–0855H = paper supplies
0870–08A8H = telephone data.

Both of these programs have 08 as the high byte of the address. This value would be stored in the interrupt vector register. The low bytes 00 and 70 are supplied by the sensing device. As soon as the telephone data is corrupted it is noticed by a sensor, and it sends a signal to interrupt the microprocessor together with the 70H.

It then combines the 08H from the I register with the 70H from the external device and puts it into the program counter. The micro-processor program then switches to address 0870H and calls the telephone maintenance engineer.

When the microprocessor is interrupted, it stores information inter-nally about what it was doing at the time of the interrupt so that when the interrupt is dealt with it can return to its previous task.

Interrupt priorities

Not all problems are treated equally, by us or by a microprocessor. It is unlikely that we would ever meet a sentence like: 'I have forgotten to put any sugar in the coffee – I'll go and get some – and I have also noticed that your house is on fire'.

As with all other microprocessor, Z80180 interrupts are partly generated by external circuits and some result from internal sources. The external ones can usually be blocked or 'masked' so we can tell the microprocessor to ignore them and all interrupts are placed in order of priority so multiple interrupts are prioritized.

In the case of the Z80180, an unrecognized instruction code is given top priority to prevent any random operation due to corruption during programming or transmission. This is called a TRAP.

After this comes a single non-maskable interrupt or NMI. This is used for critical situations that must interrupt any other program that is running.

Then follows three levels of external but preventable (maskable) interrupts. If we don't want the program interrupted, all we have to do is to insert a 'don't interrupt' code into the software. Some documentation refers to this type of interrupts as INT0, INT1, INT2, where the figure show the priority whereas as others use IRQ to stand for Interrupt ReQuest.

The remaining seven levels refer to internal interrupts generated by interrupt control registers and internal circuitry may need to interrupt the program for a moment in order to perform some other task.

There will be a little more on interrupts in Chapter 17.

Power saving

The power consumed by a microprocessor is mainly the result of internal activity so the more we make it do, the more power it uses. Changing the clock speed from 6 MHz to 33 MHz nearly doubles the power requirement.

We can add the 'sleep' instruction to the program and this has the effect of stopping the CPU clock and the data and address buses are disconnected. To wake the CPU up, we apply a signal to one of the interrupt inputs or activate the reset circuit by holding the voltage on the reset pin low.

We can also switch off the input/output circuitry on the microprocessor putting it into a 'system stop' mode.

The Z8X180 family has three members: Z80180, Z8S180 and the Z8L180. The S version operates with a power supply of 5 volts and the L version supply is reduced to 3.3 volts.

The effect on the power consumption of switching between normal operation and 'system stop' mode is shown in Table 8.1.

Bus state controller

In a system, the microprocessor is connected to all the surrounding circuits by a series of connections called the system bus. Sometimes

Table 8.1

	Normal operation	*'System Stop' mode*
Z80180	100 mW	25 mW
Z8S180	90 mW	15 mW
Z8L180	60 mW	6 mW

these devices may wish to send data along the system bus and the bus state controller ensures that we don't have multiple devices trying to send data along the same connections at the same time.

Direct memory access (DMA) controller

As the name may suggest, this circuit provides direct access to the memory without using the central processing unit. The DMA can provide high-speed transfer of data from one part of the memory to another. It can also pass data directly to and from the memory and external devices. This provides significant improvements in operating speed of the whole system.

Dynamic memory refresh

Do you remember that the problem with dynamic memories was the constant refreshing that was needed to keep the RAM data intact? Well, an internal register is used to handle the refreshing process for us. It fits in the job between instructions and does not slow the microprocessor in any way and neither the programmer nor the user is aware that it is continuing in the background. Some microprocessors do not have this register and an external chip is added to the system to perform the refreshing. Either way, it has no obvious effect on the operation of the system.

Wait states

To make it easier to send information to relatively slow external devices the microprocessor is able to insert 'wait states' which can insert more time into its bus cycle timing. They can be programmed in by the software being used or internally by a wait state generator.

Serial inputs and outputs

To move data in or out in serial mode in which the stream of data is applied to a single input connection one bit after another we are given a choice of two options.

First is a clocked serial input/output (CSIO). This is a simple high-speed data connection to another microprocessor that is capable of sending and receiving data, though not at the same time. The transmission is synchronized to the microprocessor clock.

There are also two asynchronous serial communications interfaces (ASCI). These provide two other data connections that can be programmed to select the required speed of transmission and provide two-way transmission.

How the system works

In the beginning

After the power supplies are first switched on, there is a short delay built in to allow the voltages to settle and for the clock to start. This delay is produced by a circuit similar to that shown in Figure 8.9.

Figure 8.9

The ON/OFF switch

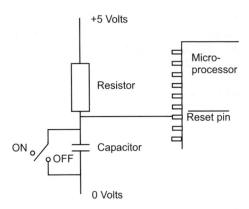

The circuit uses a capacitor. A capacitor is a device designed to store electricity rather like a bucket can be filled with water.

When the micro system is switched off, the capacitor has a short circuit across it and the current that is flowing through the resistor bypasses the capacitor hence there is no electricity stored and hence there is no voltage across it. The same effect is achieved by holding the reset pin at zero volts for a time equivalent to at least six clock cycles.

Then we switch on and the current flowing through the resistor now accumulates in the capacitor and the voltage starts to increase. The voltage grows in the way shown in Figure 8.10 and after a short period of time, less than a microsecond, the voltage reaches two volts, which is enough to switch the microprocessor on.

The clock starts ticking and the microprocessor follows a sequence of steps called a start-up microprogram that was built into the microprocessor by the manufacturer.

It performs some internal tests and similar housekeeping jobs and then puts an address on the address bus. This address is, reasonably enough, called the startup address and is again fixed by the manufacturer. In the Z80180 the startup address happens to be 0000H and so this part of the memory map must contain some ROM to hold the start-up program.

Figure 8.10

Generating a short
delay when
switching on

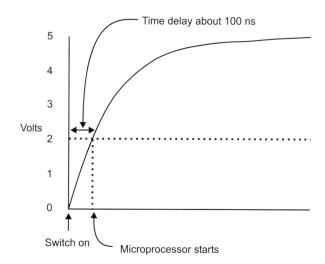

Back at Figure 6.22 we saw a typical memory map in which the start-up program was held in the high end of the memory. The Z80180 would require a quite different map and this indicates yet another incompatibility between microprocessors.

The first instruction

The startup address is read by the address decoder. The decoder then applies a chip select signal on the Control bus. This signal then selects a ROM chip. All other chips ignore the signal since they have not been selected. The address enters the ROM chip and the row and address decoders access a memory cell. The contents of that cell are put on the data bus to be read by the microprocessor.

The second read

The microprocessor sends out the next address onto the address bus. The address decoder will activate the ROM chip again. The ROM chip will put some more data on the data bus, which will again be read by the microprocessor.

What does the microprocessor do with all this data?

If the first read of the ROM resulted in the data C0H and the second read produced 86H, the microprocessor combines the two to produce a 16-bit address, C086H. Remember that all this will be in binary – the use of hex is just to help us see what is happening.

The microprocessor places this new address C086H on the address bus. The choice of C086H is not made by the microprocessor designer.

117

This number is chosen by the system designer – that is, the person that incorporates the microprocessor into a microsystem.

Then what?

The address, C086H in our example, accesses a ROM chip, which sends a binary number back to the microprocessor along the data bus. Inside the microprocessor, this number is interpreted as an instruction. Comparing the incoming number with an internal list of built-in control codes discovers exactly what the instruction is. This is all performed by the micro-program.

What happens next will depend on what the instruction was.

Why?

Why was it an instruction? Because the microprocessor said so! Microprocessors assume that the first binary number that arrives represents an instruction so if we wanted to give it the instruction ADD 25H then the binary code meaning ADD will go in first, followed by the data to be used, in this case 25H.

What if we made a mistake and put the 25 in first, followed by the binary code for ADD in our program?

The microprocessor would interpret the 25 as an instruction – which could mean anything at all, or nothing. If this mystery instruction needed a number, it would use the binary equivalent of our ADD input – which again could be anything. If the false instruction did not need a number, then ADD instruction would be correctly read as an ADD instruction but it would then take the next available binary input as the number to be added. Our microprocessor has now carried out an incorrect instruction using incorrect data and the program could do almost anything. The whole program has got out of step and may do something quite unexpected. Not much fun in our dynamite factory.

The address being used at any time is known to the microprocessor by referring to the current value in the program counter, which is increased by one every time an instruction or some data is used.

The fetch–execute cycle

This is the order of operations by a microprocessor and is the cause of all the confusion in the last paragraph.

The microprocessor applies no intelligence at all. It follows the pattern shown in Figure 8.11 regardless of whether it is following the program or it has been fed with a program with an error in it and is now carrying out a totally useless set of random instructions.

Figure 8.11

The fetch–execute cycle

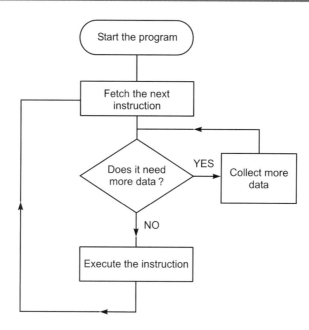

It will follow our instructions so don't blame the microprocessor – it's up to us to feed it with something sensible. Always remember GIGO – garbage in, garbage out.

As we have seen, there is nothing inherently different between an instruction and data. They are both binary numbers and the interpretation is only a matter of what the microprocessor is expecting at that particular time.

Quiz time 8

In each case, choose the best option.

1 The stack is:

(a) an abbreviation for stack pointer.
(b) a series of RAM locations that can be used by the micro-processor to store data.
(c) a collection of programs.
(d) a chimney.

2 A CPU:

(a) is an essential part of any microprocessor.
(b) is an abbreviation for computer processing unit.
(c) can contain more than one MPU.
(d) is a function of a register.

3 Adding the binary numbers 1100 1100 to 0010 1001 within a Z80180 microprocessor would result in C and H flags being in the condition:

(a) C clear and H clear.
(b) C clear and H set.
(c) C set and H clear.
(d) C set and H set.

4 The interrupt vector register (I) in the Z80180 microprocessor is used to store the:

(a) high byte of the interrupt address.
(b) start up address for the microprocessor.
(c) low byte of the interrupt address.
(d) stack pointer address.

5 The fetch–execute cycle:

(a) works in hexadecimal.
(b) assumes the second, fourth, sixth etc. inputs are data.
(c) assumes the first input is an instruction.
(d) is a system used by the Instruction register to channel information to the correct part of the microprocessor.

9

Programming – using machine code and assembly

We can use a microprocessor as part of a computer or as part of a dynamite process controller and the only essential difference is in the instructions that it is given.

Since the microprocessor has no intelligence at all, it relies entirely on following a sequence of instructions as we discussed in the fetch–execute cycle.

If you were sitting in a lecture theatre and the speaker said, 'We'll finish now', you would know that it was time to pick up your bits and pieces and leave the room. Those words were not the only possibility, other instructions would have served the same purpose like, 'Right, that's it, thank you'; 'We'll break here'; 'It seems a good time for a break'; 'We'll stop at this point and continue at the next session', and many other variations. Teachers can clear a whole classroom instantly just by looking at their watch and saying, 'Well, . . .'. There are literally dozens of ways of telling people that it is time to leave a room even before resorting to the less polite possibilities.

Microprocessors, on the other hand, have nothing like this degree of flexibility, in fact they have almost no flexibility at all. It is very frustrating, but we do not share any common language with a microprocessor (Figure 9.1). Each microprocessor has a built-in list of

Figure 9.1

A small
communication
problem

instructions that it can understand. This list is called its 'instruction set' and may consist of about a hundred or so instructions which must be put together in the right order to carry out the function required. This is the job of the programmer and is similar to a builder, constructing a house by putting a lot of simple pieces like wood, tiles, bricks in the correct order. When the microprocessor is designed, the instruction decoder recognizes these inputs and starts an internal process that allows the microprocessor to carry out the instruction. This pre-supposes that the microprocessor is familiar with the instruction or, to put it another way, the instruction is in a language that is understood by that particular microprocessor.

Figure 9.2

They are
all saying
'ADD'

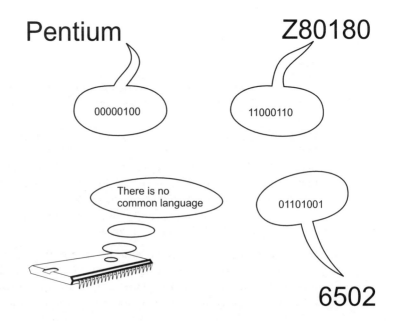

As we have seen, we cannot talk directly to a microprocessor and, even worse, microprocessors often cannot talk directly to each other (Figure 9.2). This is particularly the case when the microprocessors have been developed by rival organizations.

Within one company, such as Intel, there is a commercial pressure to ensure that each succeeding microprocessor understands the binary codes of the previous designs. This is referred to as being 'upward compatible'. There is nothing to prevent a company from designing a microprocessor that has the same pins and programming capability as another. The Z80180 for instance was designed as an updated copy of the Intel Z80 which, in turn, was a revised version of the 8080A. It was a pin for pin compatible plug-in replacement. This may be irritating for the original designers but is accepted provided the internal design has not been copied. Indeed, it often does the original company no harm. If several compatible microprocessors are being sold it will induce many programmers to write programs using this code. This will increase the sales of these microprocessors and, perhaps, no-one will suffer.

The Intel Pentium was under similar attack during 1997/8 from other microprocessors like the Athlon series made by A.M.D. These can run Pentium programs and, for some purposes, are superior to the Pentium.

Machine code

The binary code that is understood by the microprocessor is called machine code and consists of streams of binary bits. They are fed from the RAM or ROM memory chips in blocks of 8, 16, 32 or 64 depending on the microprocessor in use. To us the binary stream is total gibberish.

Example

If we refer to the Z80180 block diagram in Figure 9.3 we can investigate the instruction necessary to add two numbers together. One of the numbers is already stored in the accumulator or register A. Let's assume this is the number 25H.

In comes the instruction: 11000110 00010101. It is in two parts, the first byte, 11000110, means add the following number to the number stored in the accumulator. This first byte which contains the instruction is called the operation code, usually abbreviated to 'op code'. The second byte, 00010101, is the number 15H. This particular instruction has two bytes. Some instructions have only the one byte and others three or more bytes. The additional bytes contain the data to be used by the op code and is called the operand.

Figure 9.3

The microprocesser
adds two numbers

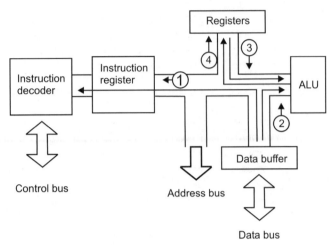

And here's the action

1 The first byte goes into the instruction decoder where it is decoded into a sequence of internal operations.

2 It then copies the number 15H from the data buffer into ALU, the arithmetic and logic unit.

3 The number 25H from the accumulator is then copied into the ALU to join the 15H which is already there. The two numbers are added.

4 And the result is copied back to the accumulator.

What is the result?

It is 3AH. Be careful not to let your brain jump back to decimal mode and shout 40 at you.

At the end of a machine-code program, we must include an instruction to stop the program, otherwise we can get some unexpected or unwanted results. You may remember that, at least in the development stages, the program is often stored in RAM.

Now, RAM locations take up random values when they are switched off so when the program has completed the last instruction, it will start executing the random values as if they were a program. These instructions may, of course, do anything at all. They could even delete or change the program that we have just written. Overall, the effect is like an aircraft overshooting the end of a runway.

The problems with machine code

There are so many. The program is not friendly: 11000110 00010101 hardly compares with 'Add 15H to the number 25H' for easy understanding. There is nothing about 11000110 which reminds us of its meaning 'add the following number to the number stored in the accumulator' so a program would need to be laboriously decoded byte by byte.

Typing in streams of ones and zeros is so boring that we will make many mistakes, particularly when we remember that a real program may be ten thousand times longer than this. Can you imagine typing in half a million bits, finding the program does not run correctly and then settling down to look for the mistakes?

Another problem is that the programmer must be aware of the internal structure of the microprocessor. How else could you know which register to use, or even which registers exist? So you master all this and then change to another microprocessor and then what? The whole learning process has to start again – new instructions, new registers, and new coding requirements. It's all too horrible.

The difficulties with machine code hardly mattered in the early days of the microprocessor. Everyone who programmed them were fanatics and loved the complexity and there were few serious jobs for the microprocessor to do. This first program language was called a 'low-level' language to differentiate it from our own verbal communication language which was called a high-level language. Machine code was later referred to as the 'First generation' language (see Figure 9.4).

Very soon, the microprocessor was used for an increasing range of tasks and revolutionary ideas like 'speed and ease' crept into the

Figure 9.4

High- and low-level languages

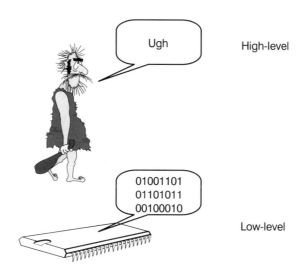

discussions. This resulted in a new language called Assembly which overcame the most immediate failings of machine code.

Assembly language, the second generation language

Assembly language was designed to do the same work as machine code but be much easier to use. It replaced all the ones and zeros with letters that were easier to remember but it is still a low-level language.

The assembly equivalent of our machine code example 11000110 00010101 is the code ADD A,m. This means 'add any nuMber ,m to the value stored in the accumulator. We can see immediately that it would be far easier to guess the meaning of ADD A,m than 11000110 00010101 and so it makes programming much easier. If we had to choose letters to represent the 'add' command, ADD A,m was obviously a good choice. A big improvement over alternatives like XYZ k,g or ABC r,h. The code ADD A,m is called a mnemonic.

A mnemonic (pronounced as nemonic) is just an aid to memory and is used for all assembly codes. Here are a couple of examples:

SLA E for shift to the left, the contents of register E.
LD B 25H load the B register with the number 25H.

Now see how easy it makes it by guessing the meaning of these:

INC H
LD C 48H

Finally, have a go at one that we have not considered yet.

LD B B'

If SLA E means shift the contents of register E one place to the left, then SRA E means shift the bits one place to the right. LD C 48H means put the number 48H into the C register. LD B B' enables us to copy the number stored in the B' register into the B register.

Note: the actual mnemonics differ between microprocessors. The manufacturers issue an 'instruction set' that lists all the codes for each of their microprocessors. Together with the number of clock cycles taken by each instruction and a summary of the function of each.

Non-destructive readout

Incidentally, microprocessors in common with memories, always use non-destructive readouts. This means that information is shifted from one place to another by copying it and leaving the original number unaltered. For example, after the instruction LD A C, the registers A and C will both finish up with the same information in them. This enables stored information to be used over and over again.

To allow us to type in ADD A,15 rather than 11000110 00010101, we need another program to do the conversion. This program is called an assembler (see Figure 9.5).

Figure 9.5

An assembler

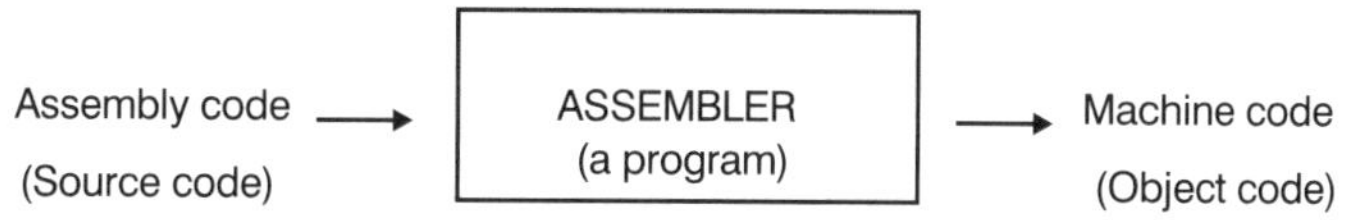

Assembly code ⟶ ASSEMBLER ⟶ Machine code
(Source code)   (a program)   (Object code)

The program allows us to type in the assembly code, called the source code, and converts it to machine code referred to as object code. It can show the object code on a monitor screen or print it out or it can load it into the RAM ready for use. When starting the assembler, we have to state the RAM starting address that we wish to use. This is normally only a matter of making sure it is in RAM and avoiding the other programs already installed. The object code is shown in hex numbers rather than binary to make it easier for us. An assembler can only work within the instruction set provided by the microprocessor designer. It cannot add any new instructions and is (almost) just a simple converter or translator between mnemonics and machine code.

Assemblers are available from many sources and all provide the necessary conversion from source code to object code. In addition, they may provide other features that will help us in the programming.

Syntax help

Syntax is the structure of statements in a language, whether it be English or a computer language. In English, most people would recognize something is incorrect about saying 'He are going' rather than 'He is going'. This is an example of a syntax error.

As an example, If we mistyped the instruction LD A C as LD A V then the assembler would be unable to convert this to object code since it will not recognize the 'V' as a valid register. A real cheapie assembler may just stop or miss out this instruction. A somewhat better one may put a message on the screen saying 'syntax error code not recognized' and a very helpful one may suggest a likely cause of the trouble. It may say, 'syntax error. Invalid register. Register name may only be A, B, C, D , E, H or L.'

Labels

Another facility offered by a good assembler is the label. A label is a word that can be used to represent an address while the program is being written.

You will recall that we have to instruct the microprocessor to stop at the end of a program otherwise it will go off following random instructions. One way of stopping the microprocessor is to give it something quite meaningless to do. Suppose you opened an envelope, and the message inside read 'put the envelope on the desk, pick it up, open the envelope and obey the instructions'. So you open the envelope, read the instructions, put it on the desk, pick it up and read the instructions. So you open the envelope, read the instructions . . . and I'm sure you can guess the next bit. We, of course, would soon stop doing this because we would see that it is pointless. A microprocessor on the other hand has no memory of ever having seen the instruction before and will be quite happy to do it forever if required.

Every byte of a program is stored in a sequence of memory locations so if we started a program in address 0300H we may have got to the address 0950H when the stop instruction is required. The last line of the program was 'jump to address 0950H'. The jump instruction tells the microprocessor to go to the address that follows, in this case 0950H, and perform the instruction to be found there. The micro-processor now finds itself in the endless loop shown in Figure 9.6. Once the microprocessor is in one of these loops it will run continuously until the power supply is switched off, or the reset pin is taken low, or one of the interrupt pins is activated.

Figure 9.6

One way of stopping a microprocessor

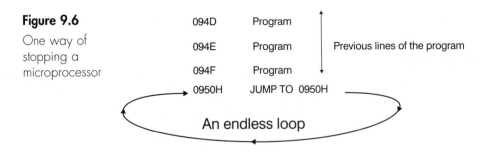

094D	Program	
094E	Program	Previous lines of the program
094F	Program	
0950H	JUMP TO 0950H	

An endless loop

So where does the label come in?

Let's imagine that we find that we have to add an extra byte at the start of the program. This will result in each byte being shifted one place down in the memory and the last line of the program actually starting at address 0951H rather than 0950H. When the program tells the microprocessor to jump to address 0950H it will no longer be in the

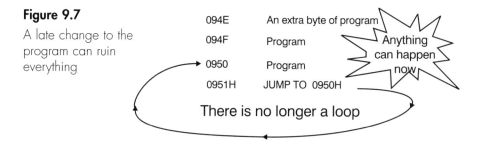

Figure 9.7

A late change to the program can ruin everything

little loop. The new contents of address 0950H may provide any random instruction and the whole program could crash. This problem, shown in Figure 9.7, will not be seen as a mistake by the assembler. Remember that the assembler is only checking for known codes, not sensible programs.

If we type a word over the position where the memory address normally resides it will be recognized by the assembler as a label. That is a word which is equivalent to the address it replaced. If we use the same label anywhere else in the program it will be given the same value. The good thing about this is that if the program is changed and the addresses change, then the value of the label is changed automatically so the two 'STOP's in the last line will always have the same value and the loop will be safe from accidental destruction (Figure 9.8).

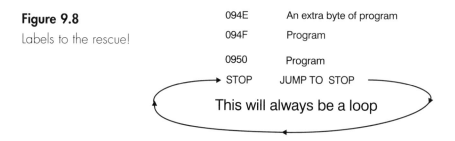

Figure 9.8

Labels to the rescue!

Once the assembler accepts the label, it can be used as often as we like in the program but remember that a label can only ever have one meaning within a single program. The choice of label is usually restricted to avoid clashing with words that have a special meaning to the assembler. Words that can do this are called 'reserved' words and usually include words such as jump, LD, ADD, HALT etc. The actual list is provided with the assembler program.

Remarks

Assemblers and all higher languages allow us to write remarks or notes to act as reminders but to be ignored by the assembler. To tell the assembler program not to attempt converting it to object code we precede the note by a semicolon or the word REM: or something similar.

The last line of our program, using the mnemonic JP for jump could be written as:

STOP   JP   STOP ;this will hold the microprocessor in a loop.

Being able to add remarks to a program makes it much easier for the program to be understood at a later date. At the time of coding, the program seems, to the programmer at least, a model of clarity. When the programmer is off sick and we have to take over, their strategy is not obvious at all. It is even more embarrassing if the program that we cannot understand is the one that we wrote ourselves a few weeks earlier. The moral of this story is to add notes even if it seems ridiculously obvious at the time.

Summary of assembly language

Assembly and machine code are not portable. This means that they are designed to be used on a particular microprocessor and are generally not able to be used on another type. They also require the programmer to have knowledge of the internal layout or architecture of the microprocessor.

Despite the two problems of portability and architecture knowledge, assembly language has survived the onslaught of the new, modern 'improved' languages considered in the next chapter.

Why?

Assembly languages have two overriding advantages in the hands of a competent programmer (note the 'competent'). A program written in Assembly is faster and is more compact, i.e. it takes less memory space to store it. Machine code and assembly languages are called procedural languages. This means that the program instructs the microprocessor to complete the first instruction, then start the next, then the next and so on until it has finished the job. This is just like a recipe.

Nearly all microprocessor-based systems are designed to operate in this way and it seems so obvious that it is difficult to think that there is any alternative – but there is, as we will see later.

Quiz time 9

In each case, choose the best option.

1   An assembler:

(a) converts assembly programs into machine code.
(b) is a type of microprocessor.
(c) converts object code into machine code.
(d) is essential for converting mnemonics into source code.

2   The data that follows the op code:

(a) is always present and consists of one or more bytes.
(b) is called the object code.
(c) is called the operand.
(d) uses denary numbers.

3   A label is:

(a) an important feature in designer clothing.
(b) a form of syntax error.
(c) used to represent an address while the program is being written.
(d) the part of the program that comes before an operand.

4   Machine code is:

(a) not a low-level language.
(b) written using mnemonics.
(c) any language designed for controlling machinery in industry.
(d) an object code.

5   The part of the microprocessor that can follow the machine code is called the:

(a) Assembler.
(b) Instruction decoder.
(c) Instruction register.
(d) Mechanic.

10

High-level languages

Third-generation languages

The third-generation languages were intended to make life easier. They were designed to improve the readability by using English words which would make it easier to understand and to sort out any faults (bugs) in the program. The process of removing bugs is called debugging. In addition, they should relieve the programmer of any need to understand the internal architecture of the microprocessor and so the program should be totally portable. Ideally the programmer should not even need to know what processor is being used. These languages are called 'high-level' and are all procedural.

Over the years, many languages have been invented just as there have been many microprocessors. Just like the microprocessors, a few languages had some special aptitude that made them stand out from the crowd. We will introduce some of the survivors.

Fortran

In the early days of computers, they were seen as a means of improving the speed and accuracy of performing mathematical calculations – rather as new and improved calculating machines. IBM dominated the computer world at that time and employed John Backus to produce an improved language to supersede assembly language.

The result, finalized in 1957, was Fortran. This was the first high-level language to gain widespread acceptance. Its claim to fame was that it could evaluate mathematical formulas. This gave rise to its name 'FORmula TRANslation' (originally 'IBM Mathematical Formula Translating System'). (See Figure 10.1.)

Figure 10.1

The first successful high-level language – and still going

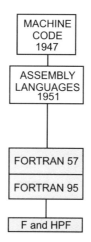

Fortran has instructions built-in to handle most scientific formulas such as finding the sine of an angle which would be extremely difficult to do in assembly language. Difficult, but not impossible. After all, the Fortran program must first be converted into the machine code understood by the microprocessor. If the Fortran program can be converted to machine code, then it follows that the program could have been written in machine code in the first case. Its just a matter of saving an enormous amount of work.

As time went by, small additions and alterations were made to the language giving rise to new 'dialects'. The disadvantage of this is that it starts to erode portability – one of the primary reasons for having a high-level language. In 1958 a re-defined language called Fortran 2 was born which was replaced in turn by Fortran 3 and 4. Still new dialects were sprouting and, in 1966, a final, last, definitive version called Fortran 66 was designed. Then we had the even more final and more definitive Fortran 77 and the absolutely final and totally definitive Fortran 90. This is not as bad as it seems since each new version included extras rather than changes. There was a Fortran 95 version and there is another due out somewhere around the year 2004 to 2006 which, at the moment, is just referred to as 200X but at least this name suggests a launch no later than the year 2009. In the meantime, an easily learnt version, simply called F, was launched. This was compatible with Fortran 77, 90 and 95 but includes some more modern characteristics while older more redundant sections were

133

quietly dropped. High Performance Fortran (HPF) is similar to the 90 and 95 versions except that it has been extended to allow micro-processors to be run in parallel for extra speed.

Compilers

In assembly language, we used an assembler program to convert the mnemonics to machine code. We usually refer to the conversions being from source code to object code but it means the same thing. In Fortran, or any high-level language, we use a compiler to produce the machine code. The compiler will also carry out the useful extras like error and syntax checking that we met with the assemblers. Compilers and assemblers are both software – that is, they are programs designed to do a specific job. If we were using a 68000 microprocessor, and wished to program it using a particular language, say Fortran, then we would have to purchase a 'Fortran 90 to 68000' compiler. It would do just this job and nothing else. We could not adapt it in any way to accept a different high-level language or 'target' it at a different processor (see Figure 10.2).

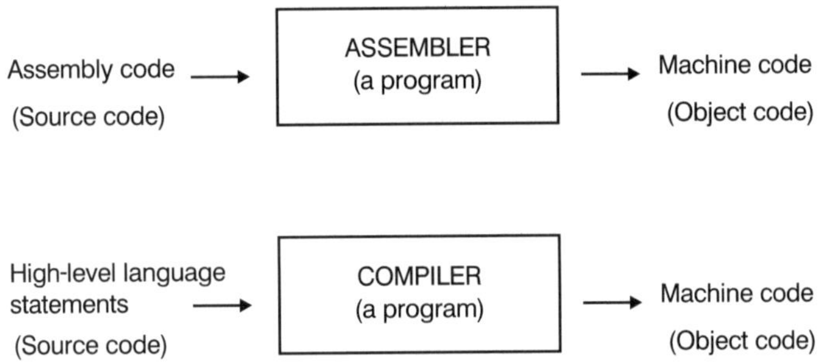

Figure 10.2

Compilers are like assemblers

Assembly code → ASSEMBLER (a program) → Machine code
(Source code)   (Object code)

High-level language statements → COMPILER (a program) → Machine code
(Source code)   (Object code)

All assemblers are basically one-to-one converters. We enter the op code and out comes the object code so the only difference between two assemblers is in the amount of debugging and label handling help that it can provide. Compilers are different. There are not any direct object code equivalents to things like sine or square root so the final output depends on the skills of the compiler designer. The programmer must look at each of the available commands in the high level language and write an assembly program to perform the function. Some compilers are, therefore, inevitably better than others.

Libraries, linkers and loaders

When the designer has struggled through the process of devising assembly code for a particularly nasty formula it would make sense to store the answer away to allow its use on another occasion. A collection of these solutions is called a library and can be purchased.

Figure 10.3

From high level to low level

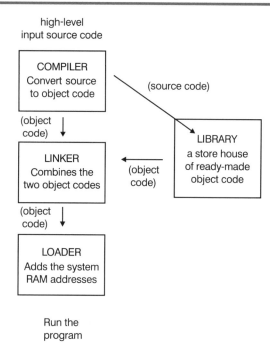

This reduces the amount of totally original coding that is needed. Many 'new' formulas can be made from combinations of existing code. Slotting these ready-made library routines into the main programs is performed by a linker which is another piece of software. The linker therefore joins or links together many separate pieces of code into one program ready for use. The last job to be done is to load it into some RAM ready for use. Another piece of software is used to determine which addresses in the microprocessor system memory are available. This is called a loader. A loader also converts labels to their final addresses. Very often the linking and loading functions are combined into a single linker-loader program. The process is illustrated in Figure 10.3.

Fortran source code

Fortran consists of numbered statements called program lines and are used to tell the system the order in which the instructions are to be carried out. In the absence of any other commands, the program lines are executed in numerical order.

Fortran code is written in a very compact form much closer to mathematics than English. For example, to load a number and find the square root of it may look like this:

```
1   Read (4) P
2   A = SQRT(P)
```

In statement 1, the microprocessor is sent to input number 4 (this must be defined earlier) and retrieves a number which we call P. Statement 2 finds its square root.

Fortran has accumulated enormous libraries to handle scientific and engineering problems. The drawback of Fortran is that its instructions are so very compact that, unless you are happy with formulas, it can look a little frightening. In addition, its format is very precise and this makes it difficult and unforgiving to learn. If you have a morbid dread of mathematics, you may find its approach a little daunting.

Basic

In Dartmouth College, USA, a simplified language was developed. It was based on Fortran and was designed as a simpler language and easier to learn. This language was called Basic (Beginners' All purpose Symbolic Instruction Code) and first appeared in 1960 (see Figure 10.4).

In the early days, the emphasis was on 'easy to learn' and 'using a minimum amount of memory' as memory was very expensive. These two attributes made it very useful in colleges but was largely ignored 'in the real world'. As microprocessors appeared and gave rise to the

Figure 10.4

The Basics

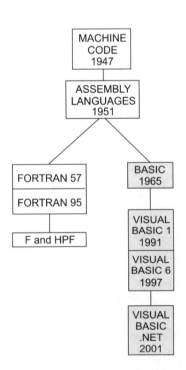

microcomputer, the benefits of low memory requirements gave it a renewed popularity and it 'took off'.

To save memory, Basic was designed as an interpreted language. An interpreter rather than a compiler carried out the conversion of the source code to the object code.

What's the difference? The compiler converts the whole program into object code, stores it, then runs the program. The interpreter takes a different approach. The first instruction in the program is converted to source code and it is then executed. The next item of source code is then converted to object code and then ran and so on right through the program (see Figure 10.5). The interpreter never stores the whole of the machine code program. It just generates it, a line at a time, as needed.

Figure 10.5

The Slow 'n' easy interpreter

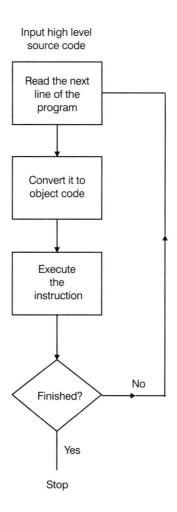

The development of Basic

The explosion of microcomputers in the 1980s resulted in widespread adoption of Basic. It was used, or at least played with, by more people than any programming language before or since. More variations, or dialects, started to appear, as occurred with Fortran.

In the case of Fortran, the American National Standard Institute (ANSI) collected all the ideas together and produced the standard Fortrans like Fortran 66, 77 and 90. This did not happen with Basic and the result is now an open market with several hundred different competing, sort of compatible, Basics floating around. Most of the earlier ones have withered away leaving a few popular versions like Q-Basic, GW Basic and Quick Basic. More recently, with the virtual monopoly of the Windows operating system, a new version called Visual Basic (VB) has appeared which has features to make the generation of Windows programs much easier.

Over the years, the language has developed to provide more and more features until it closely rivals Fortran, even for calculations. Basic has shaken off its beginner's label and is used by many professional programmers. Now that memory size is not such a problem, compiled versions are now used to accelerate the execution of programs.

There has been a series of versions of Visual Basic, VB1 to VB6 and we now have VB.Net to allow easy manipulation of Windows™ and Web site material.

Basic is very readable

It uses line numbering as in Fortran and is generally readable. Here is an example. Test out its claim to readability by guessing the outcome.

```
10   input A, B
20   Let C = A * B
30   Print C
40   End
```

On line 10, it requests two numbers A and B to be entered, possibly via a keyboard. Line 20 defines a new number C as the result of multiplying them together. Print means providing an output either onto a printer or to the screen of the monitor. The last line stops the program.

As you will remember, we always have to give a microprocessor something safe to do after it has completed a program otherwise it will start following random instructions. We can do this by sending it in a loop as we did in the assembly example to follow the same instruction over and over again or we can send it off to follow another program. The END instruction does just this. The program is returned to a 'monitor' program. A monitor program does the often forgotten parts of the system like scanning the keyboard to accept further instructions and is in no way connected with the screen showing the visual images.

The line numbers are executed in numerical order and any missing numbers are just ignored. This allows us to use line numbers that increase in steps of five or ten instead of ones. The advantage of this is that any forgotten instruction can be added later by giving it a new line number. For example, if we remembered that we wanted to divide the value of C by 2 then print this result as well, we could add a couple of extra lines:

```
10   input A, B
20   Let C = A * B
30   Print C
32   Let D = C/2
35   Print D
40   End
```

A final point is that we do not say what A, B, C or D stand for before we start. The program implies the necessary information. In other words, we do not have to declare the variables.

Cobol

Fortran and Basic, the son of Fortran, did not make enormous steps towards employing normal English language phrases. This was attempted by the US Defense Department who introduced Cobol in 1959 – just before Basic (see Figure 10.6). Its purpose was not number crunching, like Fortran, but information handling. It proved to be successful at this and spread from the US Navy where it kept records of stocks and supplies, to the business world. The name was derived from COmmon Business Oriented Language.

Cobol was designed, more in hope than reality, to be easily read by non-programmers. However, it looked friendlier than the mathematical approach of Fortran and has survived to the present day. It is generally used by large corporate computers rather than Desktop PCs.

Large businesses handle enormous amounts of information data. Just imagine the amount of information involved in a few everyday activities. We apply to open a bank account – a form appears asking an almost infinite number of questions. Then we want a credit card – more forms, more information mostly the same as we gave them for the bank account. Then we buy something. What we bought, its stock number, its price, the date, our card number, and

Figure 10.6

This is the business

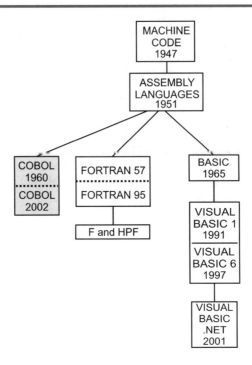

name are all transmitted to the national card centre and our account is amended. None of these transactions involve particularly complicated mathematics. The calculations are basically addition and subtraction of totals. So the calculating ability of Cobol does not need to rival Fortran or even Basic. But what it can do it to extract related information – put in our post code and out comes all sorts of information about us – credit rating, employment, home address, hobbies, purchasing patterns and almost anything else they want. Some of this information is bought and sold between companies without reference to us.

Like Fortran, it has survived by meeting a specific need and has had a series of upgraded standard versions. They refer to the date of adoption: Cobol 60, Cobol 74, Cobol 85, Cobol 97 and the new 2002 version.

A real effort was made to allow the language to be understood by those who know more about English than about programming. To this end, Cobol statements are English phrases including a verb and ending with a comma. The phrases can be joined up to form a sentence or a single statement can end with a period (full stop).

A line may read:

    1   Add staff to customers to give total people.

Pascal

Pascal was first designed in Switzerland in 1971 (Figure 10.7). It is mostly an academic language and has been largely overtaken for professional programming by languages such as C. When learning other languages, a short course of Pascal is often employed as an introduction. Pascal is used because it is 'good for you', just as it is often said that to learn European languages it is 'good' to learn Latin first to lay down the rules of language before starting on French, German or Spanish.

Figure 10.7

Pascal does you good

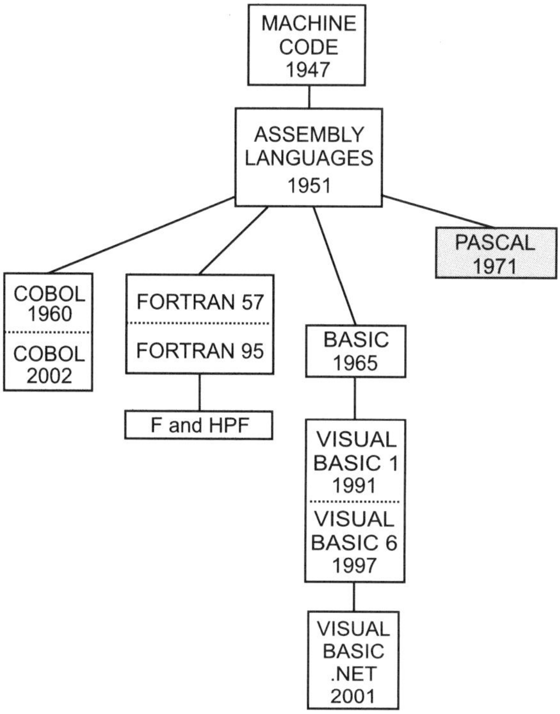

Pascal is a very structured language. A structured program consists of a series of separate, self-contained units each having a single starting point and a single exit point. The program layout looks like a simple block diagram with all the blocks arranged one under the other. Since every unit can be isolated from the ones above and below, detecting an error, or understanding a new program, is relatively easy.

Languages like Basic can use instructions like GO TO to jump to a new part of the program and this often results in what is called 'spaghetti' programming, making it very difficult to find a fault in the program or

141

Figure 10.8

Structure
and
spaghetti

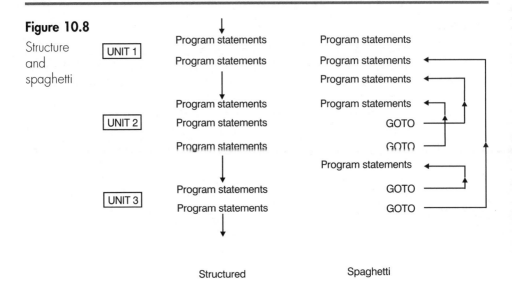

Structured   Spaghetti

even to understand what the program does (see Figure 10.8). Pascal avoids this by using instructions like 'Repeat ... until'. Basic is cleaning up its act by incorporating this type of instruction into the more recent versions.

C

Apart from teaching good programming habits, Pascal was largely replaced by the language called C, invented a year after Pascal and allowing all the good practice programming methods of Pascal with a few extras (see Figure 10.9).

The main difference is that it is a lower-level language than Pascal which may seem a strange improvement. Its advantage is that it can control low-level features like memory loading that we last met in assembly without all the drawbacks of using assembly language. It has many high-level features, and low-level facilities when we require them and can produce very compact, and therefore fast, code.

C++ and object-oriented programming

A new version of C has included all of the previous C language and added a new feature called object-oriented programming. This version, called C++, is referred to as a superset of C (see Figure 10.10).

Object-oriented programming is a somewhat different approach to programming. In all previous cases, a task has been set and we look at

Figure 10.9

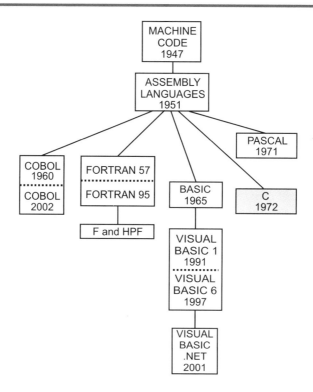

the problem in terms of what processing is required to reach the required result. In object-oriented languages we have a number of objects which can be anything from some data, a diagram on a monitor screen, a block of text or a complete program. Once we have defined our objects, we can then allocate them to their own storage areas and define ways of acting on the entire object at the same time.

As an example, if we drew a square on the monitor screen and wished to move it, we could approach this in two ways. We could take each point on the screen and shift its position and hence rebuild the square in a different position. The object-oriented approach would be to define the shape as an object, then instruct the object to move. This is rather similar to our way of handling parts of the screen in a Windows environment. We use a mouse to take hold of an object, say a menu, and simply drag it to a new position. The menu is being treated as a single lump, which is an example of an object.

All the menus have similarities and differences. Similar objects are grouped into 'classes'. A class includes the definition and type of the data together with the methods of manipulation that can be applied to an object. In our example of a class including the menus, each specific

143

Figure 10.10

C plus objects and
Java

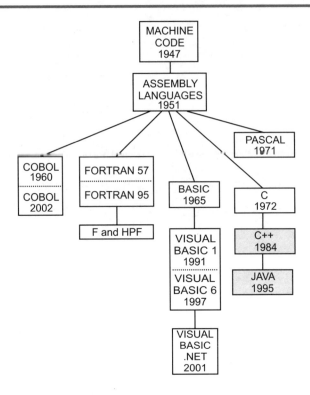

menu is called an 'instance'. All of a class share some properties called 'public' properties and are different in some way, like different text being entered, these are called 'private' properties.

As we saw earlier, C++ is just standard C with the object-oriented extras added to it.

Java

Although not its original destiny, Java was found to be ideal for transmitting information over the Internet. It looks, at first glance, to be similar to C++ but it has some important advantages. It is small and does not require any particular architecture and can therefore be embedded in other applications for use in a wide range of systems. This embedded Java code is called 'Applets' and is used in Internet Explorer and other browsers.

Fourth-generation languages

Fourth-generation languages are non-procedural and tend to concentrate on what the program must do rather than the mechanics of the step-by-step approach of the procedural languages. A tempting definition is to say they are the most recent, or the most popular or the 'best' languages. None of these definitions apply, so it may be better to stick to non-procedural as the most likely definition.

Lisp

Lisp was designed about the same time as Cobol and several years before Basic. It is the work of an American, John McCarthy (see Figure 10.11).

Lisp (LISt Processing) involves the manipulation of data which is held in lists or entered by keyboard and is associated with artificial intelligence. Lisp is a function-oriented language. This means that we define a function, such as add, subtract or more complex combinations. A list consists of a series of 'members' separated by spaces and enclosed in brackets. Samples of lists are (2 5 56 234) or (mother father son daughter).

A simple function defined as (PLUS 6 4) would return the answer 10 by adding the two numbers. Since it is an interpreted language, the program is executed one step at a time so inputted values are used as they are entered. In some versions of Lisp, this would have been written as (+ 6 4) using the mathematical symbol.

Figure 10.11

Lisp – dealing with lists

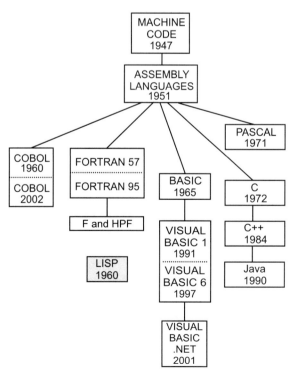

We can define our own function by saying

>(defun result (A B)(+A B))

defun = define function and A B are inputted numbers and the answers have been given the name 'result'. So, if we input the numbers 4 and 5 we would receive the response 9. This has defined a function in which we enter two numbers A and B and they are added. We could use this to enter a list of values for A and B generating a list of all the results.

APL

The letters APL stand, reasonably enough, for 'A Programming Language'. This is another interpreted language developed by IBM around 1962 and is only used for numerical data. It is a curious mixture of Lisp and Fortran. It combines the function orientation of Lisp with the terse procedural mathematics of Fortran (see Figure 10.12). It allows user-defined functions and has a large library of solutions to common problems. Most people would agree that given a choice, APL is not the language to learn if you are in a hurry. For example, the four basic functions of add, subtract, multiply and

Figure 10.12

A programming language

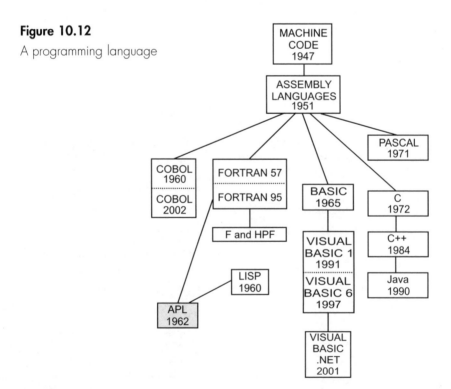

divide are present but even here, life is not easy. What would you expect the result of 2 * 3 to be? Well, it's not 6. This would be written as 3 × 2 and 2 * 3 is actually 2^3 or 8. The statement: Value ← 4 – ⁻2 four take away minus two giving an answer of plus six. Notice the different symbols for minus and a negative number. Other mathematical functions like sin, cos and tan are replaced by special symbols that do not appear on standard keyboards.

Easy, it is not. The good news is that, once mastered, it provides fast compact programs but the many cryptic statements would need to include many comments to help another person to understand your program.

Prolog

Prolog is called a 'declarative' language in which the program designer does not need to know exactly what the output will be when starting the design of the program. It was first developed in France in 1972 with a view to its use in the development of artificial intelligence. Prolog stands for PROgramming by LOGic. Other versions were developed, such as DEC10, IC Prolog, which were produced in the UK and other versions from the US (see Figure 10.13).

Figure 10.13

A logic-based language

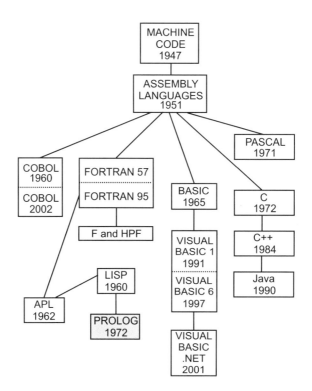

Prolog is another non-procedural language in that it is not a route to a goal but a set of information and a method from which the result can be deduced. Basically the idea is to feed in some facts and ask the program to produce some conclusions. You will remember the little logic puzzles like 'Graham is married to Anne, Kirk is the son of Peter and the brother of Matt . . . Is Kirk the brother of . . .' You know the sort of thing – more and more interconnected pieces of information until your head hurts. Just the job for Prolog.

The program includes facts and rules then we can ask questions. Here is a really simple example.

Facts:
coins (franc, france)
coins (centime, france)
coins (dollar, usa)
coins (cent, usa)

Rules:
french(x):–coins (x,france)
american(x):–coins (x,american)

Now we ask some questions:

?french (centime)
answer: Yes

?french (dollar)
answer: No

OK so far, but:

?american (dime)
answer: No

It has no data so it cannot say that dime is correct so it plays safe and says it is incorrect.

The future

Increasing microprocessor speeds and using several to share the processing tasks together with the decreasing size and cost of memory will be the key to the future. The idea of a desktop computer running at 20 GHz and having 128 Gbytes of memory is no longer ridiculous. In fact, it is looking rather modest after looking at Table 10.1 in which the trend over the last 26 years is projected another 26 years into the future.

'It cannot keep increasing' – quote of the year 1972, 1973, 1974, 1975, 1976 . . .

Table 10.1

Similarly priced microprocessor systems	Intel 4004 in 1972	Intel Pentium in 1998	Microprocessor in 2024
Clock speed	0.108 MHz	300 MHz	833 GHz
Memory	640 bytes	64 Mbytes	6.4 Tbytes
Performance (MIPS)	0.06 approx.	600 approx.	6 000 000

A persistent occupant of my crystal ball is real spoken voice communication. A few years ago voice recognition was only a dream and is now a reality and becoming increasingly efficient. Voice synthesis is progressing nicely and is beginning to sound less robotic. When these two technologies mature, simultaneous language translation will not be far away and real dialog with the computer will begin. Another million television programs suddenly become available without sub-titles (you see, all progress comes at a price!).

Six million, million instructions per second and 800 GHz clock speed together with total voice control of computer functions will be here in a few years.

'How about that Mr Spock?'
'Fascinating'

That's my guess.

At least it is something to look back and smile about in the future.

Quiz time 10

In each case, choose the best option.

1   A compiler:

(a) converts machine code to a high level language.
(b) is faster than an interpreter.
(c) is another name for a linker.
(d) is not available for the Basic language.

2   APL was largely influenced by:

(a) Cobol and Prolog.
(b) Lisp and Fortran.
(c) Fortran and Basic.
(d) Pascal and Cobol.

149

3 **A fourth generation language can be described as a language which:**

(a) is still being used.
(b) is object-oriented.
(c) was developed for artificial intelligence.
(d) is non-procedural.

4 **A language designed to allow logical deductions to be made from input data is:**

(a) C.
(b) Latin.
(c) Prolog.
(d) Fortran.

5 **Pascal:**

(a) is a low level language compared with C.
(b) is only of use if you are going to translate it to European languages.
(c) is a highly structured language.
(d) was the first popular high-level language.

11

The development of microprocessors and microcontrollers

Micros are getting bigger – and faster

As the complexity of microprocessors and other digital integrated circuits has increased, there has been an inevitable increase in the number of transistors that are incorporated in their design.

In the list below, we have used transistors or their equivalent. These classifications are not universally accepted, there are different names and numbers floating around, so a degree of flexibility should be employed when comparing different sources. This is particularly true at the large end where the terminology has not yet 'firmed up'.

SSI	Small scale integration	1–10 transistors
MSI	Medium scale integration	10–1000 transistors
LSI	Large scale integration	1000–10 000 transistors
VLSI	Very large scale integration	10 000–100 000 transistors
SLSI	Super large scale integration	100 000–1 million transistors
ULSI	Ultra large scale integration	1–10 million transistors

The increase in the number of devices has also had the effect of necessarily decreasing the size of each component. If the same component size were used for the current front runners as was used for the original 4004 microprocessor, they would be about the same size as a page of this book. For reasons that we will look at in a moment, unless we reduced the size of the components, we couldn't increase the speed of operation and so the current microprocessors would have a maximum clock speed of under 1 MHz.

How do we measure the speed of a microprocessor?

This is a lot more difficult than we think because the developers of microprocessors are in competition with each other so as soon as a method is suggested, they try to exploit the situation to present their microprocessor as faster than all the others.

Be wary when reading comparisons – which tests have they chosen, and why? While watching the last Olympic games, it occurred to me that I was probably faster that any of those competing in the 100 metres. Yes, I felt confident that I could build a working micro-processor-based system quicker than any of the athletes. You see, comparisons all depend on the test that we have decided to use. Anyone can be World champion. It's only a matter of choosing the tests well enough. With that in mind, here are a few popular speed comparisons.

MIPS (millions of instructions per second)

This appears an easy measurement to take. It is simply a matter of multiplying the number of clock cycles in a second by the clock cycles taken to complete an instruction.

The current Athlon for example can run at 2 GHz or 2000 MHz. It can perform up to 9 instructions per clock cycle so its number of instructions per second is simply $2 \times 10^9 \times 9 = 18\,000$ million cycles per second.

Life is never that simple. Some instructions are more time consuming than others as they take a different number of clock cycles to perform the task. Competitors will obviously choose instructions that give the most impressive results on their own microprocessor.

An extreme example occurred about ten years ago with the Intel 80836, we could ask it to perform some additions. By consulting the instruction set, we could see that they each take two clock cycles to complete. Now, if we take a clock frequency of 25 MHz, each clock cycle would last for 40 ns so an 'add' instruction would take 80 ns. This would equate to a speed of 12.5 MIPS. An unkind competitor may take 'at random' the divide instruction that takes 46 clock cycles, they could reduce its MIPS rating to a measly 0.54 MIPS. A really vindictive person could search through the instruction set with a magnifying glass and find that there is a really obscure instruction that takes 316 clock cycles. This would provide a speed of 0.08 MIPS – about the same as a four-bit microprocessor.

We cannot even say that we can make it fair by using the same instruction for each microprocessor since they don't always have their talents in the same area. If we had two microprocessors, each with a 'load' instruction taking five clock cycles and they both ran on a 10 MHz clock, what would be their speed? 10/5 = 2 MIPS. (In working this out we can just ignore the fact that the clock cycle is in megahertz since the speed is measured in millions of instructions.) But can they do the job at the same speed? Possibly or possibly not. What if one was a 64-bit microprocessor and the other was an 8-bit microprocessor. The 64-bit one could shovel data at eight times the speed. For good reason MIPS have been referred to as 'Meaningless Indication of Performance by Sales reps'.

FLOPS (FLoating-point Operations Per Second)

To overcome the problem of which instructions should be employed some standard floating-point operations can be used.

As a quick reminder, a floating point number is one in which we have moved the decimal point to the start of the number, so 123.456 would be converted to 1.23456×10^2. This makes the mathematics faster. Modern microprocessors would have values in the order of ten GFLOPS (GigaFLOPS).

This also meets with objections. The obvious question is, 'What "operation" is being measured?' Choose your operation carefully and the opposition is left far behind.

Both of these tests, MIPS and FLOPS are supposed to be microprocessor tests and not system tests.

System tests

Other speed measurements tend to be system tests rather than microprocessor tests but a brief overview may be in order since they are often quoted, almost as alternatives.

Benchmarks

These tests are based on making the microprocessor-based system run a standard or 'benchmark' program. The immediate failing here is that to get a program to run on microprocessors with incompatible code will mean that the compilers will also be tested, which is not part of the system. There is also an immediate outcry from people who disagree with the program chosen since it doesn't suit their system.

153

I/O operations (input/output operations)

As the name suggests, this measures the speed of accepting information in and sending it out again. But loading information from a CD for example will depend on whether the information is being read from the same track or does the head have to move and include seek-time?

TPS (Transactions Per Second)

This is a move to model the tasks set into real-life situations. It requires the system to take in information, modify it and then store it again. It puts a heavy significance on memory access times and compilers.

SPECmark (Systems Performance Evaluation Co-operative's benchmark)

This is the average result of carrying out 10 agreed benchmark tests in an attempt to measure the system performance in a range of situations. Recent changes include one using floating point arithmetic, which is of more interest to serious number crunching in science and engineering and the other an integer test for the rest of us. These are referred to as SPECfp95 and SPECint95. As a starting point for comparison, the 200 MHz Intel Pentium Pro delivers a value of 8.71 using the SPECint95.

How to make a microprocessor go faster?

Increase the clock speed

This seems the obvious answer and generating the necessary square wave shown in Figure 6.4 is no problem. In a modern microprocessor-based system there are two clocks that we need to consider. There is a square-wave clock that controls the internal operation of the microprocessor. This is the headline speed seen in the adverts, 'the 2 GHz Pentium'. There is also the operating clock, about 133 MHz, for the system to control the external devices and memory. This saves us from upgrading all the external devices to match each new processor.

Internal clock speeds will probably continue to increase at least into the low gigahertz range but there are limiting factors that will make continuous increases in speed difficult to achieve. It would be a trivial electronic problem to generate a square wave of 1000 GHz or more, so there is obviously more to it. And there is.

Power dissipation

Heat is a form of power and is an unwanted by-product of any activity inside a microprocessor.

Power = voltage × current

So, to reduce heat production, we have to reduce either the voltage or the current, or both.

We will start with voltage since it is a little more straightforward. The early microprocessors used a 15 V supply that has been steadily reduced and the latest designs are pushing at 1.5 V. How much further can we go along this line? It appears as if we have nearly reached the limit. The integrated circuits are made from a semiconductor called silicon. This passes electricity under the control of electric charges. The applied voltage creates this charge. In silicon, the simplest device needs at least 0.6 V to operate although, by adding minute traces of other materials, this figure can be reduced a little. A single transistor can do little with voltages less than 1 V so in a complex circuit it is already amazing that the total voltages can be as low as they are. The chances of a microprocessor running on less than 1 V is slight indeed. If I were braver, I would say impossible. Another point with regard to the voltage is that we must not forget the effects of random electrical noise as we saw earlier in Figure 2.2. Sudden changes in current flow in nearby circuits can cause random changes. If the voltages are reduced too far, the microprocessor will become more prone to random errors.

Bursts of current are promoted by the vertical leading and trailing edges of the clock pulses so the higher the clock frequency, the more edges per second and the more current will flow and hence more heat will be generated. To reduce the heat generated simply reduce the clock speed, which is exactly what we don't want to do.

Size of architecture and its effects

As the electric charges move through the transistors inside the microprocessor it takes a finite time. It follows then, that if we reduce the size of the transistors we can move data around faster and this is true.

The smallest feature that could be fabricated in a microprocessor had an initial size of about 10 μm when microprocessors were first produced; it has now been reduced to 0.13 μm, a significant reduction. This reduction has two drawbacks. Firstly, it is much more difficult and expensive to manufacture without accepting enormous failure rates. Secondly, the heat generated has not changed, since it is a feature of voltage and current but not size. This means that its temperature will increase unless we can dissipate the power. An unfortunate problem with semiconductors is that they are heat sensitive and will auto-destruct if the temperature rises too far. We do our best with heat sinks, which are basically slabs of aluminum with fins to increase the surface area, and fans to keep the heat moving. A typical operating range is 0–85°C when measured in the centre of the outer case of the microprocessor (not the heat sink).

155

To increase the system speed

When we look at the overall system, it is apparent that not all things have progressed at the same rate. The largest bottleneck is the memory. We already use a slower clock speed but the microprocessor still spends a lot of its time humming a tune and bending paper clips waiting for information to arrive from the memory. During the life of the microprocessor, the clock speed has increased from 0.1 MHz to about 3 GHz, an increase of about 30 000 times. During this time, these DRAM memories have got bigger but only about 2000 times faster.

Modern microprocessors have about 128 kbyte of on-board RAM, called a cache. When the microprocessor has to go to the external memory for information, it saves a copy of the address and the information in case it is needed again. It also saves the address and information from the next memory location. The reasoning behind this is that since nearly all languages are procedural then the next location is likely to be accessed next. If not, the program may jump back to a previous address to repeat part of a program as in a counting loop to produce a delay. When the microprocessor next requires access to the memory, it first checks the high speed cache to see if the information is stored, if it is, we have scored a 'hit' and the system has increased its speed. If it is not there, it is a 'miss' and the main memory is used. This new information is then stored in the cache for later.

This cache is sometimes called a level 1 cache, or L1 cache. This implies that there may be a level 2 cache – and there is. The L2 cache is usually 256 kbyte.

When data is needed, the microprocessor checks cache level 1, then 2 and lastly, the main memory.

Making more use of each clock pulse

Pipelining

To put too much reliance on the clock frequency is like saying that the maximum rpm of the engine determines the maximum speed of a vehicle. Yes, true, but other things like gearbox ratios are also significant. Doing 9000 rpm in first gear will not break any speed records. The real speed of a microprocessor also depends on how much useful work is done during each clock cycle. This is where pipelining is really helpful and is now incorporated in all microprocessors.

Let's assume we have some numbers to move from the memory to the arithmetic and logic unit (ALU):

Clock pulse 1   A number is moved from a memory location to the accumulator.

Clock pulse 2   It is then moved from the accumulator to the ALU.

If we have another number to be loaded, this would have to repeat the process so loading two numbers would take four clock pulses. Three numbers would take six clock pulses and so on.

During the first clock pulse, a number is being moved along the bus between the memory and the accumulator and so the other part of the bus between the accumulator and the ALU is not used. During the second pulse, we still have one section of the bus idle (Figure 11.1).

Figure 11.1

One clock pulse moves one number

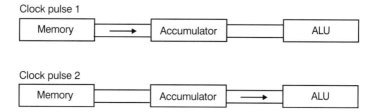

Pipelining is the process of making better use of the buses. While one number is shifted from the memory to the accumulator, we can use the same clock pulse to shift another number from the accumulator into the ALU along the other section of the bus. In this way, we get more action for each clock pulse and so the microprocessor completes instructions faster without an increase in the clock speed (Figure 11.2).

Figure 11.2

One clock pulse moves two numbers

If we get two jobs being done on the same clock cycle, then this has made a significant improvement to the speed without increasing the clock speed. If we can manage to get three pieces of information moving, or jobs done, this is even better. Incidentally the Pentium manages five and the Pentium Pro can manage 12 and the Pentium 4 can keep up to 126 instructions 'in flight'. Unfortunately, we can never get pipelining to work this well on all instructions, but every little helps.

RISC and CISC

If we wished to AND two binary numbers, we could do it by using a logic gate as we saw in Chapter 5 or we could use a microprocessor executing an instruction code. Now, comparing middle-of-the-range devices, the logic gate would complete the task in 8 ns but a comparable microprocessor (80386, 25 MHz) would take a minimum of 80 ns.

This type of comparison established the belief that, given a choice, hardware is always faster than software. In the above case, it is 10 times faster.

Given the job of carrying out a hundred such instructions we had a choice.

Software method = 100 operations × 80 ns = 8000 ns (8 ms)
Hardware method = 1 operation at, say, 240 ns + 100 hardware
operations × 8 ns = 1049 ns

This philosophy was followed throughout the development of 4- and 8-bit microprocessors. This gave rise to more complex hardware and a steady increase in the size of the instruction set from a little under 50 instructions for the 4004 up to nearly 250 in the case of the Pentium Pro.

In the mid 1980s, the hardware-for-speed approach began to be questioned. The ever-increasing number and complexity of the operating codes was reversed in some designs. These microprocessors were called RISC (Reduced Instruction Set Computers) and the 'old fashioned' designs were dubbed CISC or (Complex Instruction Set Computers). History has not proved so black and white as this suggests. It is much more a matter of shades of grey with new designs being neither wholly CISC nor RISC. The use of predominately CISC microprocessors outnumbers RISC designs by a wide margin, at least 60:1. This does not imply that they are better but they simply have a greater proportion of the market. As we know, there is a lot more to market dominance than having the best product. Sadly.

CISC designs include all the 8-bit microprocessors and Pentium, Pentium Pro and all of the 68000 family whereas the RISC includes Digital Alphas and the IBM/Motorola Power PCs.

RISC versus CISC

Both RISC and CISC microprocessors employ all the go-faster techniques such as pipelining, superscalar structures and caches. A superscalar architecture is when there are two ALUs that share the processing like having two microprocessors. So, what are the real differences?

By analysing the code actually produced by compilers, we find that a small number of different instructions account for a very large proportion of the object code produced. Most popular are the instructions that deal with data being moved around.

158

At this point a curious switch of design occurred. You will remember that the 'normal' or CISC microprocessor included a microprogram in its instruction decoder or control unit. This microprogram was responsible for the internal steps necessary to carry out the instructions in the instruction code. So the microprocessor that we have been praising for its use of hardware to gain speed, is actually being run internally by software.

The RISC approach was to reduce the number of instructions available but keep them simple and do them fast. The number of instructions were reduced to under a hundred. Since instruction codes can be easily enhanced by adding some extras to the microprogram it was tempting to do it. No pruning of previous instruction was possible owing to the need to maintain compatibility with previous versions.

Following the cries of 'hardware is faster than software' it seemed a logical step to do away with the microprogram and replace it with hardware that could carry out the simple steps necessary. This hardware was made more simple by keeping all the instructions the same length so that pipelining was easier to organize. The only disadvantage of these constant length instructions is that they all have to be the same length as the longest and so the total program length will be increased.

CISC kept shoveling bucketfuls of data backwards and forwards between the microprocessor and the external memory using many different types of instruction. RISC designs just had a simple load and store instruction and everything else is done internally using a large number of registers to replace the external memory.

By the use of hardware for handling instructions and internal moves between registers, all instructions could be reduced to a single clock cycle, which gave a significant increase in speed. Generally, the Pentium Pro has managed to match this speed by its extensive use of pipelining.

As time goes by, there is an increasing tendency for the RISC/CISC difference to decrease. Modern RISC microprocessors like the PowerPC970 have an increasing number of instructions even though they do tend to be simple and fast and the more traditional CISC approach in the Pentium 4 is also employing simple yet extremely fast instructions.

Who did what, when

As we know, our starting point was the Intel 4004 in 1972, very quickly followed by the 8-bit 8008 processor. These are shown in Figure 11.3. Notice that, even within a company, there was no agreement about the operating voltages.

159

Figure 11.3

The first 4- and 8-bit microprocessors

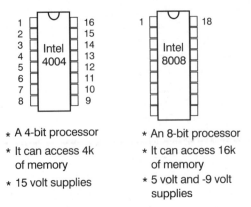

* A 4-bit processor
* It can access 4k of memory
* 15 volt supplies

* An 8-bit processor
* It can access 16k of memory
* 5 volt and -9 volt supplies

It also started the trend for numbering, rather than naming, a microprocessor. This made good sense since each basic design generated a series of variations, different speeds, modified instruction sets etc. The numbers can give a clue as to some of the basic characteristics and the hierarchy. Sometimes an X is used to signify a family of devices like the 80X86, so by giving different values to the X, we include the 80286, 80386 etc. The tendency now is to use a name and a number. This was due more to a legal problem owing to the difficulties of 'owning' a number. This was highlighted after Intel produced the 80286 followed by the 80386 and 80486, which gave its competitors rather advanced warning of the name of the next one. Intel couldn't copyright the number 80586 otherwise mathematicians would fear prosecution if any calculation resulted in this number. They tried to call it the P5, claiming world rights over the letter P. Finally they went for 'Pentium'. Apart from the fun of watching it all going on, the main beneficiaries, as usual, were the corporate lawyers (for whom we all pay, of course).

Intel versus Motorola

In December 1973, Intel introduced the 8080A. This was a very popular processor that had a 16-bit address bus so it can address up to 64 kbytes of memory or, as the adverts said at the time, 'a MASSIVE 64k of memory'. The power supplies changed again, this time to +5 V, +12 V and −5 V. The number of instructions increased again and the number of pins increased to 40. Internally, there was the normal accumulator and eight general-purpose registers. This, then, became the standard package size for future 8-bit microprocessors as shown in Figure 11.4.

At about the same time, Motorola introduced a rival in the form of the MC6800. Like the Intel 8080A, this one was an 8-bit microprocessor

Figure 11.4

A standard size for 8-bit
microprocessors

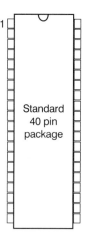

with 16 address pins. At this point, the similarities ended. It was not compatible with the 8080A and so fought toe to toe in the market place, and came second!

The power supplies were simplified, now requiring only a single +5 V supply. A block diagram of the MC6800 is shown in Figure 11.5. We can see that it was unusual in not having any general purpose registers but it did have two accumulators. The approach here was to use an external memory location for the temporary storage of information that all previous microprocessors would have put into an internal register. The remainder of the microprocessor is quite familiar from our look at the Z80 in Chapter 8. The 6800 performed slightly faster than the 8080 on average, but not enough to break the hold of Intel in the marketplace.

The final 8-bit microprocessors

At this time, Intel was working on the design of a replacement for the 8080A. It was a response to the criticisms of the 8080A: Why does it need three different power supplies when the 6800 only need one? And why only one interrupt pin when the 6800 has two?

At this time some of the engineers that had been working on the 8080A and were developing the replacement called the 8085A decided, or were persuaded, to move to a rival company called Zilog.

Meanwhile, back at Intel, the 8085 was produced. It answered all the gripes about the 8080A and increased the clock speed to 5.5 MHz. Still 8 bits with a 16-bit address bus, it stayed with its eight internal general purpose registers. It followed Motorola's lead and opted for a single +5 V power supply but to keep its customer base it kept its instruction code compatible with the 8080A.

161

Figure 11.5

MC6800 Motorola's answer
to the Intel 8080A

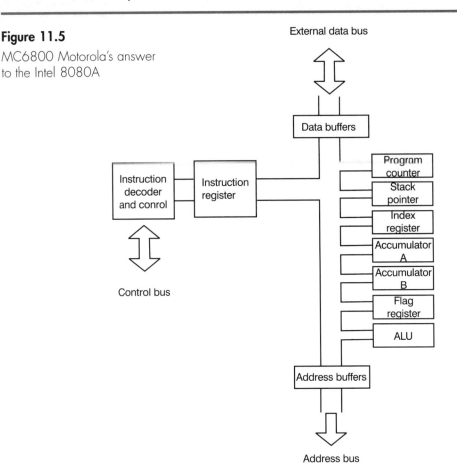

Back at Zilog, the other group of engineers that was also brought up with the 8080A set to work on the Z80 which was later developed into the Z80180. They combined much that was good about the Intel designs with some good ideas from the MC6800. A single power supply was used and this not only returned to the use of internal general purpose registers but increased their number to 14. The instruction code included all the code from the 8080A but added some new ones to nearly double the total number.

Meanwhile a new player, MOS Technology entered the fray with its own MCS650X family of which the MCS6502 is probably the best known. This flew to fame with the rise of the microcomputer in the 1980s. It was basically an enhancement of the Motorola MC6800 and follows the 8-bit trend of a 16-bit address bus and a single +5 V power supply. Its contribution to progress was the idea of pipelining. Two billion 6502s were sold.

Returning to the plot

The 6502 was said to be 'accumulator-based' in that it has no general purpose registers and a single accumulator through which all the incoming and outgoing data is passed but with pipelining and some fast instructions – it was very popular for a few years. In comparing Figure 11.6 with Figure 11.5 we can see the influence of the MC6800. Most of the blocks are the same. The MC6800 had two accumulators and one index register and the MCS6502 has two index registers and one accumulator.

Figure 11.6

MCS6502 – an enhancement
of the MC6800

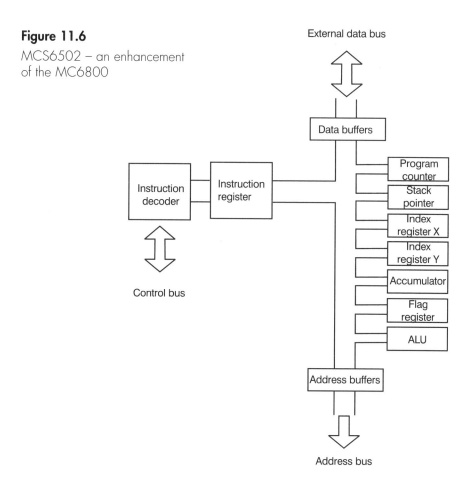

There was little to choose between these microprocessors, as the 6502 had the edge in the microcomputer market while the MC6800 was more popular in the industrial control field.

The one-chip microcomputer

To make a stand alone system, the microprocessor would require some RAM and ROM memory as we saw in Figure 7.6. In many industrial control situations the size of the memory would not need to be very large and it occurred to the designers that all these necessary parts could be built into a single chip. When this occurred the name microprocessor was superseded by the one-chip microcomputer.

Intel produced the 8048. This was not simply an 8085 + RAM + ROM, but a new design. It included (yet another) change of instruction code so it was not compatible with either the original 8080A or the 8085A. The on-board memory consisted of 1 kbyte of ROM and 64 bytes of RAM which was quickly doubled to 2 kbytes and 128 bytes on a new version, the 8049. It could also access 4 kbytes of external ROM. A further enhancement was a timer. This can count up or down to provide time delays. Without this, the microprocessor would have to be used for this function which would prevent it from getting on with something more useful.

Zilog, of course, were not far behind. It bought out the Z8. This was similar in concept but slightly upgraded. It had two counter/timers that could be used for counting incoming pulses as well as providing the time delays. It had 2 kbytes of ROM and 128 bytes of RAM that it called general purpose registers. A big improvement was its ability to access 64 kbytes of external ROM and 64 kbytes of external RAM to allow the 'one-chip' to become more than one if needed.

Motorola replied with its MC6801 that included many of the features of others in its generation. It included 2 kbytes of ROM and 128 bytes of RAM together with a timer and UART. Its instruction set was compatible with the MC6800 with a glimmer of 16-bit arithmetic creeping in.

Rockwell launched their R6500, which was basically a 6502 microprocessor with the addition of 2 kbytes of ROM and 64 bytes of

Figure 11.7

The first four generations

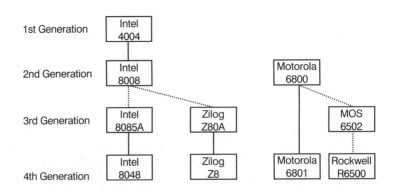

RAM. One interesting feature was that the internal RAM had a separate power supply in the form of a battery so that the data wasn't lost during a power failure. It also included a universal asynchronous receiver/transmitter (UART) that we will save for Chapter 15. The first four generations of microprocessors are shown in Figure 11.7.

The microcontroller

As the home computer was being developed, the headlines were only interested in the ever-increasing speed and capability of the micro-processors. However, behind the scenes, the microcontroller was selling in greater numbers with little publicity. There is not much in the way of attention-grabbing headlines in a microcontroller being fitted into a video recorder.

The mammoth engines of today's computers were just not required for many uses in the real world. We are the limiting factor when it comes to everyday uses of microelectronics. We may want our computers to work faster but we still work at the same speed that our ancestors. Just how fast do we want a vending machine to work? We want it to work at the same speed today as it did yesterday, or ten years ago. After all we cannot press the buttons any faster and we don't want the coffee to fly down the chute at ten time the previous speed.

If we want any change, we may like it to be more reliable. Is the Pentium 4 any more reliable than the Z80? I doubt it. We would certainly want it to be cheaper and use less power, and not require a fan if it thinks about something.

The trend is to rethink what we really want and the answer in the majority of cases is something much closer to the present day microcontroller.

We will be coming back to microcontrollers in more detail in Chapters 15 and 16.

Increasing the number of bits

It was inevitable that the 4-bit microprocessor that turned into the 8-bit should, in turn, grow into a 16-bit microprocessor but first a few moments to answer a seemingly obvious question.

What do we mean by a 16-bit microprocessor?

The 'size' of a microprocessor is the width of the data registers, so an 8-bit microprocessor can handle 8-bit numbers. It was traditional that 4- and 8-bit microprocessors had an address bus that was twice the width of the data registers. This was just a coincidence and doesn't follow these days since no one wants, or could afford, the memory to

165

fully use a 128-bit address bus (the number of locations is more than 3 followed by 38 zeros!)

The other fallacy is that it is necessarily the same as the width of the data bus. It is not. The Pentium family uses a 64-bit data bus but they have 32-bit data registers and are therefore 32-bit devices. It uses the 64-bit data bus to load two 32-bit registers at a time. The Power PC and the Digital Alpha family are real 64-bit devices and have a 128-bit data bus.

Curiously, the Intel 8088 was a real 16-bit microprocessor but had an 8-bit external data bus. This was to allow it to be compatible with the cheaper 8-bit circuitry, which had not quite caught up with the idea of using 16 bits.

Even so, the data registers are not universally accepted as defining the size of a microprocessor. Some people stick with the data bus to be the defining size. So, in reference books and catalogues you may find microprocessors referred to as a different 'size' to the one you expected. In this book I will stick with the data register as being the defining feature.

The 68000 family

The M68000, first produced in 1979 was a VLSI chip employing about 70 000 transistors. The M68000 microprocessor is well known as a 16-bit microprocessor but, in reality, it is a 32-bit device if we stick to our definition above. It certainly has a 16-bit data bus but the internal registers are 32-bit although some arithmetic operations can only use 16-bit data. Occasionally this format is called a 16/32-bit processor. It was in a 64 pin dil (dual-in-line) package, as shown in Figure 11.4, but even longer. Its length was often its undoing since it could easily snap in half if you attempted to remove it by prizing up one end of the chip. It has a 24-bit address bus that can therefore access 16 Mbytes of memory with a 12 MHz clock frequency.

One feature of M68000 is its pre-fetch action. When an instruction is being worked on, the microprocessor fetches the next instruction from memory and stores it in a little queue, ready to be used. This can be done whenever the present instruction is not using the external address and data buses. This means that the next instruction is already loaded ready to go as soon as it is required, thus saving valuable time. It has a total of seventeen 32-bit general-purpose registers of which eight are data registers, which can be used as 8-, 16-, or 32-bit registers as required.

One interesting feature is that this microprocessor can operate in two modes, supervisory and user. The essential difference is that the user mode has a restricted list of instructions at its disposal. The operating system can use the supervisory mode and thus use the full set of instructions but user programs only have access to a restricted range – enough to run the programs but, hopefully, not enough to screw things up too much. There is a software route between the two modes if you really want to change.

The M68000 gave rise to its own family as the basic model progressed. The main advances are detailed in Figure 11.8.

Figure 11.8

The M68000 family

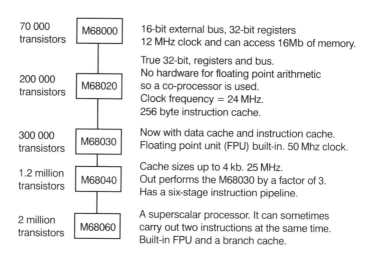

70 000 transistors	M68000	16-bit external bus, 32-bit registers 12 MHz clock and can access 16Mb of memory.
200 000 transistors	M68020	True 32-bit, registers and bus. No hardware for floating point arithmetic so a co-processor is used. Clock frequency = 24 MHz. 256 byte instruction cache.
300 000 transistors	M68030	Now with data cache and instruction cache. Floating point unit (FPU) built-in. 50 Mhz clock.
1.2 million transistors	M68040	Cache sizes up to 4 kb. 25 MHz. Out performs the M68030 by a factor of 3. Has a six-stage instruction pipeline.
2 million transistors	M68060	A superscalar processor. It can sometimes carry out two instructions at the same time. Built-in FPU and a branch cache.

There are a few new terms here.
They will be explained shortly.

As true 32- and 64-bit microprocessors have taken over the computing side, the 68000 family is used increasingly as a high-performance embedded control for printers and disk drives.

Where do we go from here?

After the early one-chip microcomputer, the decreasing cost of the design and production of the integrated circuits made it easier to increase the complexity of the chips.

This has caused the development to diverge along two separate paths: speed and power or cheap and small.

By heading off in the cheap and small route, we get microcontrollers that are controlling the operation of most of the instruments and machines that we use and even playing tunes in greetings cards. We will meet these in Chapters 15 and 16.

The never-ending pursuit of more speed and more power for computers has resulted in the continuous development of larger and faster microprocessors like the Pentium 4 and its competitors. Each new design is king for a day, and then overtaken and dispatched to the museum. What cost a fortune three years ago is thrown out with the garbage, unwanted. We will look at these here today, gone tomorrow devices in the next three chapters.

Games machines

If we only needed computers to allow us to handle text on a word processor, there would be little need for the development that has occurred in the last ten years – our typing is not getting faster and text is simple stuff. Introducing coloured pictures does little to increase the stress levels but things really start to change when we want to have moving coloured pictures.

We are never satisfied: we want faster-changing, more lifelike images – with sound effects, of course. Our demands will outstrip the latest microprocessors almost regardless of their capability.

Many computers are used mostly for playing games and simulations but because they have other functions they must be designed to operate across a wide range of fields and not really optimized for any particular task. This has resulted in the development of the dedicated games machine. We have a choice of three at the moment – they are the Nintendo Gamecube, the Playstation and the X-Box.

Nintendo Gamecube

This is the oldest design and, inevitably, the least technically advanced. It is generally cheap to buy the cube and the games are also cheap and are most popular amongst users in the lower age groups.

They decided to use a 64-bit, 85 MHz, IBM PowerPC 750Cxe microprocessor, also called the 'Gekko' which was its original codename during the production stages. It is a slightly modified version of the one used in Macintosh computers and similar to those described in Chapter 13. The modifications were software additions of almost forty additional instructions to provide specific help in game playing but not necessary in the original computer applications.

The quality of the graphics during play can be indicated by the speed of handling graphic data which is conveniently measured by how many shapes it is able to draw in one second. In this case the chosen ATI 162 MHz Flipper GPU (Graphics Processor Unit) has a maximum speed of 12 million polygons/second – this sounds fast but compared with the Playstation 2 and the X-box it is not impressive.

The main game storage is a 1.5 GB 3 inch Nintendo optical disk. There is also a facility for a 64 MB memory card supplied by Panasonic.

Sony Playstation 2

Rather than taking a ready-made microprocessor off the shelf and then designing a games machine around it, Sony started by designing their own microprocessor called the 'Emotion Engine' designed just to run games programs as fast as possible. This 'single job' approach allows the design to be very focused. The microprocessor is surrounded by the necessary circuitry to provide the necessary inputs and outputs. The overall blocks are shown in Figure 11.9.

Figure 11.9

Playstation 2

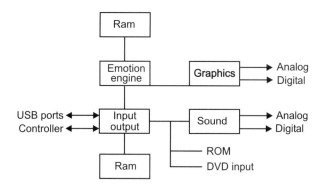

If we load a game from a DVD, the startup information is passed to the Emotion Engine which prepares the game graphics and sound in either analog or digital format. It then waits for input instructions arriving from our controller or through the USB ports.

Inside the Emotion Engine, shown in Figure 11.10, our controller information arrives through the Input/Output unit and the fun begins.

Figure 11.10

The Emotion Engine

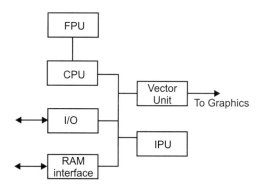

Games involve serious numbers of calculations and they can be very complex. There are two types of calculations: straightforward calculations (just the sort of thing we could do on a pocket calculator) and geometric calculations. The ordinary calculations are performed by the FPU (Floating-Point Unit) and the others are done by the VU (Vector Unit). So why so many calculations?

Does the car skid on the corner? This depends on how fast you are telling it to turn, what speed you are driving at, the weather selected and the car data. What is going to happen if you 'accidentally' hit a tree or another vehicle?

These calculations have to be done in real time – if we turn the steering wheel the car has to respond immediately. The geometric calculations are performed by the Vector unit, which provides the results of what is happening on screen. It also prepares the list of events that control everything that appears on the screen – right down to the path taken by the wheel that has broken off and the reflection in the driver's mirror.

To a large extent, the final quality of the game experience depends on the speed of these calculations. Floating point calculations are performed at more than six billion a second! That is moving.

So, how good is the Playstation 2?

That is too difficult to answer. Instead we can look at the technical information but there is much more. Does the controller feel good? Are the games exciting and realistic? Does an hour on the PS2 seem like minutes or weeks?

Generally, the console is fairly expensive but, having bought it, the games are cheap(ish). The thinking behind this is we only buy the PS2 once and soon forget how much it cost, especially if it was a gift, but we can afford plenty of games. Having plenty of games reduces the attraction of changing to another games machine – like the Xbox.

The Emotion Engine runs at only 294 MHz but these headline speeds are not a good indication of how fast it can do its job – this also depends on how well it has been designed for the job. You may remember that the Gamecube microprocessor was running at 485 MHz, more than 50% faster than the PS2, but look at the drawing speeds: Gamecube 12 million polygons/sec, PS2 25 million polygons/sec, double the speed. This compares a general-purpose computer microprocessor with a dedicated device.

The Microsoft Xbox

The Xbox is the latest offering in this market. They have opted for the opposite strategy to the Playstation 2 by making the console fairly cheap but charging more for the games. Presumably, the name of

Figure 11.11

The Xbox

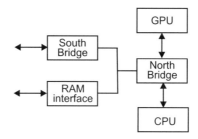

Microsoft together with an attractive upfront price will mean the boxes are carried out the shops and we will worry about the games later. They seem to sell to an older group than the other two and perhaps the price of games may not be such a barrier.

The microprocessor chosen for this machine is the Pentium 3 running at 733 MHz. They have gone for a standard micro rather than a specialized design but even so, it has got such power that it can blast through a game at a good rate. The geometric drawing speed of textured polygons is 50 million per second or twice the speed on the PS2 or eight times that of the Gamecube. In all cases, the polygon count has texture included. The bare polygon is a simple wire-frame shape without any surface coloring which is essential to provide reality to the scene.

The North Bridge chip is the central block of the Xbox and provides interconnections between the other units. It controls access to the 64 MB memory that provides a cache for the use of the CPU (Central Processing Unit) to store program code and a larger share for the GPU (Graphics Processing Unit). The North Bridge also sends signals to the South Bridge chip.

The South Bridge provides all the external inputs and outputs via the USB and network ports together with the audio signals.

The GPU design shown in Figure 11.12 is called the 'nVidia' is based on the GeForce 3 design, another popular computer component. The graphic processor converts the CPU output into the finished information being sent to the CPU or television.

The information about each object on the screen such as its position, the lighting applied and the surface appearance is prepared together with two features that are used to decrease the computing power necessary or to improve the appearance with the same processor. The

Figure 11.12

The graphics chip

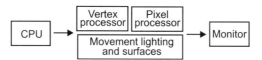

'pixel shader' that can apply realistic lighting and surface texture effects over a whole scene without each individual point being calculated separately and the vertex program oversees the detailed changes that are necessary in critical areas. We tend to be very selective when it comes to details, we would notice the slightest change in a facial expression yet ignore whole leaves on a tree. By controlling light and texture characteristics, the vertex shader fills in details of these small but important changes.

Quiz time 11

In each case, choose the best option.

1   The 'size' of a microprocessor is determined by the:

(a) width of its data registers.
(b) number of lines in its external data bus.
(c) number of digits in its type number.
(d) width of its address registers.

2   An integrated circuit having 15 000 transistors is classed as a:

(a) LSI device.
(b) SLSI device.
(c) VLSI device.
(d) SSI device.

3   An untextured polygon:

(a) looks like a dinosaur.
(b) is just a wire-frame shape.
(c) has no shape.
(d) is a cube.

4   An L1 cache is usually:

(a) onboard the microprocessor.
(b) constructed from DRAM for maximum speed.
(c) slower than Level 2 cache.
(d) external to the microprocessor.

5   RISC:

(a) means 'radical instruction set computer'.
(b) has longer instructions and is therefore slower than a CISC chip.
(c) is part of everyday life.
(d) chips employ a smaller instruction set.

12

The Pentium family

The Pentium is a 32-bit microprocessor just like the previous Intel 80386 and 80486 but has been considerably enhanced to improve its speed of operation. Even the 132 pins of the 80386 have increased to 296 on the Pentium.

Other full RISC chips were being well-received at the time the CISC Pentium was launched in 1993 and Intel took these new designs into account but it was boxed into a corner by its own success. It had to maintain absolute compatibility with the previous 8086, 80286, 80386 and the 80486 together with their numerical co-processors. The compromise was to use all the RISC while maintaining the CISC codes. It has over 400 instruction codes. Some are performed by hardware and some by microcode. Its two million plus transistors have been incorporated into a superscalar structure. This means that it has duplicated arithmetic and logic units that can allow it to carry out two instructions at the same time under favourable conditions.

It was launched at 66 MHz and in its first year became famous as the microprocessor that couldn't count. There was a flurry of letters in the computer magazines and a host of 'How many Pentiums does it take to change a light bulb?' type jokes. At first, Intel denied there was a problem even though they must have known about it. 'And, no, you can't have your money back.' More letters. 'Alright, there is a very, very small matter of a few division sums.' The error actually produced inaccuracies in the sixth or ninth decimal place in some particular division sums. This was insufficient error to affect more that a small minority of users but it started to undermine confidence in the Pentium. The real problem was that two errors occurred during its design at the same time. Either one, on its own, would have been spotted but the two mistakes served to hide each other. Anyway, it's been fixed. It only affected the early versions and is no longer significant.

Over time the speed has increased to 200 MHz with the inevitable rumours of the Pentium II running at 400 MHz that will support a 100 MHz system clock.

An outline of Pentium operation

See Figure 12.1.

Data and code caches

Connections to the outside world are via a 64-bit external data bus and a 32-bit address bus. The incoming data that consists of numerical data and instruction codes are loaded very quickly into two internal caches – an 8 kbyte data cache and an 8 kbyte code cache. These caches shift data very rapidly on the internal pathways that are 128 and 256 bits wide.

Whenever possible, the Pentium uses burst mode to read and write data. The burst mode system loads a cache for example, with more data than the width of the data bus. If a cache line is 128 bits wide and it is fed from a 64-bit data bus, then we could completely fill the line

Figure 12.1

The Pentium processor

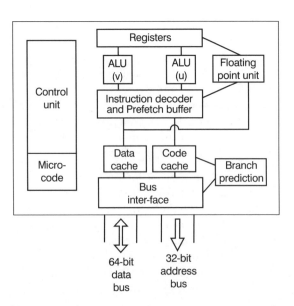

by transferring 64 bits and then another 64 bits. Burst mode loads all 128 bits very rapidly without further intervention from the micro-processor. Putting more new data into the cache will increase the chances of the cache holding the required information. This is called a cache 'hit'.

Prefetch buffer

The prefetch buffer is a small internal memory that holds a list of instructions that are waiting to be executed. This ensures that the instruction decoder is never waiting for a new instruction from the external (slow) memory and it makes more efficient use of the external data bus since the new instructions can be loaded whenever the opportunity arises. When it gets a moment, the Pentium shifts an instruction from the external program into the cache and transfers one instruction from the cache into the prefetch buffer and also sends a signal to the microcode circuit to prepare the code for the next instruction. So, with all the housekeeping done, the instruction decoder can be fed with instructions and data at its maximum rate. The prefetch buffer is actually two independent 32-bit buffers, each providing input to one of the ALUs.

The instruction decoder

The instruction decoder performs much the same function as in other microprocessors. It has two outputs that are fed to the two ALUs called 'u' and 'v'.

Arithmetic and logic units

These units are under the control of the aptly named control unit. The blocks, shown in the diagram as ALU 'u' and ALU 'v' are actually five step pipelines that can operate in parallel to execute two instructions in a single clock cycle. All commands other than floating point arithmetic can be executed in the 'u' pipeline and a more limited range can be carried out in the 'v' pipeline. The five-stage pipeline can speed the throughput to one instruction per clock cycle. In the correct conditions, both pipelines can be used simultaneously to handle two instructions in a single clock cycle. Sometimes this is not possible. Perhaps both instructions need access to the same piece of hardware, perhaps the result of an instruction is needed before the next instruction can be started. As a rather simplistic example, if we wished to add two numbers then divide the result by 10, we cannot start dividing anything until the first answer is available. One minor drawback is that instructions cannot overtake each other even if the second one could have been finished very rapidly and they are not dependent on each other.

175

Floating-point unit (FPU)

For floating-point arithmetic, the FPU has an 8-bit pipeline that is further enhanced by using a hardware multiplier and divider. This is a significant advance over the 80486, which was not pipelined in the FPU. Between them, the pipeline and the hardware, the FPU runs about ten times faster than the 80486 with equivalent clock speeds. You may remember from earlier discussions that one of the benefits of the RISC designs was the use of hardware for the execution of arithmetic operations.

The 'u' pipeline has some overlap with the floating-point pipeline so there are restrictions on the occasions when two instructions can be executed at the same time.

There are eight FPU registers 80-bits wide, arranged as a stack. Bits 0 to 63 hold a 64-bit mantissa. Bits 64 to 78 hold a 15-bit exponent and the last bit holds a sign bit.

S	Exponent	Mantissa

Bit   79 78   64   63   0

Notice how the layout of the floating-point number differs from the example that we saw in Chapter 4.

Branch prediction

When the program reaches a 'branch' or 'jump' instruction, the microprocessor is sent to another part of the program. These instructions are usually 'conditional' as in 'jump to address xxxx if the value in the accumulator is not zero'. When this jump happens, the next few instructions that are loaded into the pipelines are all incorrect and the pipeline has to be emptied and restocked with the new information. This is called 'flushing' the pipeline and causes an irritating delay of four or five clock cycles.

The branch prediction logic holds about 256 entries in a cache to aid the Pentium in guessing the next instruction. If we can guess what is coming next before it happens, then the data and instructions can be loaded ready to go.

But how do we guess? There are two likely outcomes: either the branch will be taken and we jump to another part of the program, or we don't take the branch and we continue with the next instruction. The branch prediction logic argues that what the microprocessor did last time, it will probably do again. This is true more often than not. The reasoning behind this is that when a loop occurs, the program is sent back to repeat a section several, or many, times. It can only NOT take the branch once, so on average it will take a branch more often than it doesn't.

In the cache are stored the instructions immediately before the branch or jump together with the target address assuming the branch is taken. It also stores statistical information of how often the branch was taken in the past. This information is used to predict the likely outcome of the current situation and is correct for about 85% of the time. When the branch has occurred, the history information is updated to make the next guess even better.

General purpose registers

The Pentium has seven general-purpose registers, all 32-bits wide. One of them is used as an accumulator and to maintain compatibility with the 80386 and the 80486, it can be addressed as a single 32-bit register, two 16-bit or four 8-bit registers. There are three other general-purpose registers that can be similarly split and three that only offer the choice of 32- and 16-bit use.

Interrupts

The handling of interrupts has not changed beyond all recognition since we were looking at the Z80.

There are two hardware interrupts available. The NMI or non-maskable interrupt is activated by the pin voltage going to a logic 1 or high-level. Immediately on the completion of the current instruction, the Pentium puts the content of the flag register and the current address onto the stack. It then goes to the flag register and resets the interrupt flag to prevent any further interrupts. It then services the interrupt. The NMI normally occurs as a result of hardware failures to quickly limit the damage caused.

The IRQ or interrupt request is also activated by the appropriate pin going to a logic 1 or high-level but in this case remember that it is only a request and can be blocked by resetting the interrupt flag in the flag register. If more than one interrupt is received they are checked for priority and the highest one wins. IRQs are generally initiated by peripheral equipment such as a printer.

Exceptions

These interrupts are issued by the microprocessor itself and occur when the microprocessor has found itself in a difficulty that it cannot resolve.

When an exception occurs, an on-screen message often appears announcing that an exception has occurred and the Pentium attempts

177

the instruction again. Asking the Pentium for an impossible answer causes some exceptions. This could be 'division by zero'. Dividing any number by zero is not possible and the Pentium cannot respond.

Another one, which often strikes terror into the heart of the user, is 'General Protection Error'. The software has sent the Pentium off to an address that doesn't exist and obviously, therefore, no instructions are available.

The MMX Pentium

MMX (MultiMedia eXtensions) is an addition to the standard Pentium designed to increase the speed of multimedia, communications and other applications where large numbers of repetitive calculations are required.

It started by analysing a wide range of typical applications: graphics, video, games, speech recognition etc. Intel was looking for time-consuming common characteristics. Many were found in which a fairly simple instruction like changing the colour of a pixel is applied to a large number of pixels. This gave rise to the idea called SIMD (Single Instruction Multiple Data). Using SIMD, we can perform the same operation on multiple bits of data, and this is executed in parallel. MMX allows eight pixels to be moved around and process them together. SIMD is the heart of MMX.

MMX technology maintains full compatibility with previous instructions and has added a further 57 instructions. No danger of the RISC approach here!

MMX instructions take over control of the eight floating-point registers and it has a further eight registers for holding addresses, loop control, data manipulation instructions etc. The floating-point registers are highly flexible in that the 64-bit mantissa section can be used for eight separate bytes, four 16-bit words, two 32-bit 'doublewords' or a single 64-bit 'quadword'.

Saturation arithmetic

In normal fixed-point arithmetic adding two numbers can cause an overflow to occur and the msb can be lost. To take a simple example, adding the number 1 to the byte 1111 would give the result 10000. This would offer the result of zero and an overflow would have occurred as seen in Chapter 4. To check for the overflow, the microprocessor would have to take time out to check the status register to see if the overflow flag has been set. This is time consuming. When applied to graphics, perhaps shading, the sudden return to zero may cause a sudden and unwanted change in colour.

Figure 12.2

Saturation arithmetic prevents wraparound

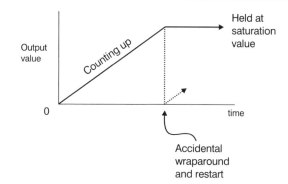

Saturation arithmetic ensures that any increase that would cause a wrap-around effect of returning the value to zero is prevented (see Figure 12.2). If we counted up from 0000, the Pentium would allow the count to proceed normally until it reached the maximum value of 1111 and it would then be held at that value. The colour in our example would reach black but would be prevented from accidentally returning to white.

The Pentium 4

As we have mentioned before, one of the limits on operational speed is the size of the internal components and, until recently the smallest detail was limited to 0.18 μm. As the competition between the AMD continued, it was time for the next step as AMD started using 0.13 μm technology and, as expected, the Pentium 4 also upgraded to the same technology for the faster versions of 1.8 GHz and above. The operating voltage has also been reduced from 1.75 down to 1.5 volts allowing closer spacing and a further increase in speed (and 25% reduction in cost). The new design has allowed the Pentium 4 to increase its transistor headcount from 42 million to 55 million increasing the number of connecting pins to 478. Intel has moved a long way from the 16 pins of their 4-bit offering in 1972.

Thermal safety

The power dissipation increases as any integrated circuit works faster and the Pentium 4 is no exception. Now, bearing in mind that the actual processor circuit is just 10 mm × 10 mm (0.4 square inches) and consumes 55 watts. We must be very careful to ensure that it doesn't overheat. This is achieved by using a large heat sink and a cooling fan. The new Pentium has a thermal safety circuit. If the microprocessor starts to overheat, the cooling fan will increase its revs and the operating speed of the microprocessor will decrease. If things get

179

serious and it reaches a dangerous level of 69°C (155°F) the thermal circuit will call it a day and shut down the computer to prevent the microprocessor from being destroyed.

The system bus

Also called the FSB or Front Side Bus, is 64 bits wide and 'Quad Pumped' which is a fancy way of saying that each clock pulse, presently running at 133 MHz, will shift four lots of data along the bus. Now, rounding off the figures a bit, 133 MHz × 4 = 533 MHz so the bus looks like a single 533 MHz bus. Incoming and outgoing information is stored in the 256 kB level 2 Advanced Transfer Cache which is fed 256 bit wide pathways. Intel calls it 'Advanced Transfer Cache' which is not quad pumped though being wider, still matches the speed of the system bus.

Instruction Decoder, Level 1 Execution Trace Cache and Branch Predictor

The data that is selected by the predictor is loaded into a buffer and then passed onto the Instruction Decoder.

At this stage, the incoming instructions are analysed and converted into an internal code sequence which can be accessed from the Micro code as we saw when we looked at the Z80180 microprocessor.

Once the instructions have been decoded, up to about 12 000 instructions called 'Micro-Operation/Operand' or μOP are stored in order of use, all ready to go. The correct order is much assisted by the Branch prediction – known by Intel as the Branch Target Buffer (BTB). This stores previous experience to guess what is likely to happen next.

Hyper pipeline

As we saw in Chapter 11, the pipeline is the organization of the microprocessor and not a separate device within the design, so we don't get a 'pipeline' block shown in Figure 12.3. The predictor designs are now very much improved, having had the experience gained with earlier versions of the Pentium. The better the prediction, the longer and faster we can risk make the pipeline. So pleased were Intel with their predictions that they called the new longer pipeline a 'hyper pipeline'. For maximum speed we would like a long pipeline so that many simple steps can be carried out at greater speed but the overall outcome depends on the predictor circuits making the right guess. A wrong guess means that the pipeline is loaded with incorrect data and has to be refilled, or 'flushed', which takes valuable time. The

Figure 12.3

The Pentium 4
processor

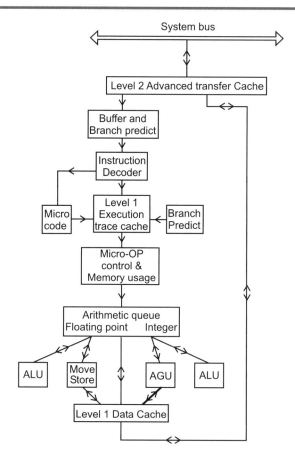

Pentium 4 now has a pipeline of 20 stages which allow 126 instructions to be in use at a single time which can include up to 48 load and 24 store instructions.

Micro-OP and Memory usage

The µOps that pour out of the Execution Trace Cache are arranged in order and they will be a mixture of information to be stored in memory locations and arithmetic operations. The arithmetic operations are divided in floating-point operations and integer operations. The floating-point register deals with moving and storing while the ALU (Arithmetic and Logic Unit) deals with the more complex operations such as multiplication of 128-bit numbers and MMX (multimedia instructions) as we met a little earlier. The SIMD (Single Instruction Multiple Data) that was applied to the earlier Pentiums have been extended by an extra 144 instructions. This facility is now called SSE2 (Streaming SIMD Extensions 2 instructions). The general idea is that if we have to perform an action on many bits of data, it is simpler and faster to collect them all together and perform the function on all of them at the same time.

Rapid Execution Engine

For the integer instructions there are two ALUs clocked at twice the core processor speed which is a four-fold improvement over the basic function and provides a data transfer rate of 48 GB/s.

A level 1 data cache handles the data outputs from the ALUs and the AGUs (Address Generation Units).

Future development

The new Pentium design with speeds over 1.8 GHz and 0.13 µm technology is given the codename 'Northwood' that replaces the previous 'Williamette'. The Williamette had reached the end of its development whereas the Northwood is just starting and since it is already running at 2.8 GHz, the magic 3 GHz chip is imminent, then we can probably look forward to the even more magic 4 GHz before the Northwood design is obsolete.

The Celeron

The bold type in computer adverts always shouts about 'price and speed' and many people fall into the trap of assuming that a 2.8 GHz microprocessor is obviously faster than a 2.5 GHz microprocessor. This is a false assumption but still well established so for this section of the market there is a demand for a very cheap microprocessor with a high clock speed.

The solution is to use the Pentium design and cheapen it by taking out some of the non-essential areas. There have been twelve such versions to track the Pentium releases during its development. In the 2 GHz Celeron, the price reduction has been achieved at the expense of reducing the L2 cache from 512 kB down to 128 kB and the FSB down from 533 MHz to 400 MHz.

Quiz time 12

In each case, choose the best option.

1   SIMD is:

(a) used in standard Pentiums but not in the MMX versions.
(b) a way of preventing wraparound.
(c) single in-line multimedia data.
(d) single instruction multiple data.

2 Branch prediction logic:

(a) is another name for the prefetch register.
(b) is only used in MMX versions.
(c) saves memory in 85% of occasions.
(d) attempts to guess the future steps to be taken by a program.

3 An exception:

(a) will be ignored if the I flag is set to a high level.
(b) is an unusual branching of the program.
(c) is an interrupt signal generated by the microprocessor.
(d) occurs whenever the Pentium is surprised by an arithmetic result.

4 The initials SIMD stand for:

(a) SIM card type D.
(b) Single Instruction Multiple Data.
(c) Superscalar Instruction Mode for Data.
(d) Streaming Instructions Modular Data.

5 In its construction, the Pentium 4 uses:

(a) 0.13 μm technology.
(b) 1.8 μm technology.
(c) 1.3 μm technology.
(d) 0.18 μm technology.

13

The PowerPC

Intel was producing a series of CISC microprocessors and, together with Microsoft, was in a position to dominate the market. Being increasingly squeezed out was the traditional king of computers, IBM, which, at one time, produced more computers that all other manufacturers combined. Big Blue as IBM was called, on account of their logo and the blue suits worn by their army of salesmen, laid down the standard design for the computer that now eclipses all others designs in the world.

As long ago as the mid-1970s IBM had developed a RISC micro-processor but it didn't really make it in the market place. RISC did not 'come of age' until Acorn produced the ARM 2 and 3 microprocessors for their Archimedes microcomputer, but this too, failed to muscle its way into the market as it made little attempt to make it compatible with Intel code. Acorn, at that time, was introducing the Archimedes as a replacement for the much-loved BBC microcomputer. By 1990, it was apparent that the terrible twins, Microsoft and Intel, would take over the world if no one fought back.

As it happens, this was the very year in which a fledgling company called 'AMD' was hatched to grow over the years to become a persistent irritant to Intel. As yet, Microsoft still rules the world but there is a system called Linux that may, one day, become troublesome.

Meanwhile, an alliance was formed between IBM, Motorola and Apple Computers. To this alliance IBM brought their POWER microprocessor (Performance Optimized With Enhanced RISC). This was the successor to the earlier 801 RISC microprocessor and was chosen because it was a RISC microprocessor and already had software developed. Motorola would build the chip and Apple would bring its computer operating system, which was light years ahead of the Microsoft equivalent at that time. The new family of micro-processors was to be called the PowerPC series.

The designers took great care to make it attractive to software companies by being careful to address the problem of future development. They distinguished between the overall architectural features that will stay the same throughout the series, rather than how these features will actually be implemented. This allows the pro-grammers to know which parts they can rely on to be consistent and which bits are likely to change. For example, they designed the system for 64-bit operation even though only 32 bits were to be used in the early devices.

The PowerPC 601 (or MPC601)

The PowerPC 601 was introduced in 1994 and followed the agreed PowerPC architecture as shown in Figure 13.1. It used 2.8 million transistors which is slightly less than the Pentium but many of the Pentium transistors were tied up with maintaining compatibility with their earlier microprocessors.

Figure 13.1

The PowerPC 601 architecture

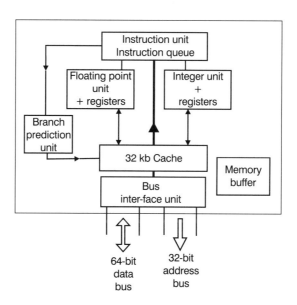

Many of the blocks shown are familiar after our look at the Pentium. The 601 is a 32-bit microprocessor using a 64-bit data bus and a 32-bit address bus.

Bus interface unit

This serves the usual purpose of connecting the data bus and address buses to the microprocessor. It also acts as a control device to determine whether the data is to be read into the microprocessor or written into the external memory.

Cache

This is a single 32 kbyte cache which is shared by data and instructions. Later versions have increased the total cache available to provide two separate 32 kbyte caches, one for data and the other for instructions.

Within the cache, the information is arranged in a series of groups or lines of 64 bytes. To provide a high-speed link between the cache, the bus interface unit and the instruction unit and queue, a 256-bit internal bus is provided.

On many occasions, the result of a particular instruction is not of great interest in itself but just provides the data to be used for a future instruction. So when an instruction is completed, the result is stored in the cache rather than being put back into the main memory. Writing the result back into the cache is called a 'write-back' organization as opposed to 'write-through' action when the information is sent to the external memory. This, of course, saves a lot of time since the cache is about seven times faster than accessing the main memory and a million times faster than using the hard drive.

Instruction queue and instruction unit

The fast internal bus maintains a queue of up to eight instructions. Using the normal RISC ideas, all the instructions are the same length at 32 bits. Eight such instructions can fit across the 256-bit width of the internal bus.

The function of the instruction unit is to send instructions to the three pipelines: integer unit, floating-point unit and the branch prediction unit. With the right mixture of instructions, we can handle three instructions at the same time. To keep the pipelines busy, it also has the facility of running some of the instruction out of order. This is limited to instructions that are not interdependent.

The branch prediction unit

In the Pentium, the branch prediction included analysis of the history of each branch or jump instruction to help predict whether it is likely to be taken. The PowerPC uses a single stage pipeline which decodes and executes in a single clock cycle employing a very much simpler strategy that curiously seems to work just as well.

It makes no choices. If the branch is sending the program back to an earlier instruction, it always assumes that the branch will be taken. This is usually the correct choice since such loops in programs are very common. On the other hand, if the branch instruction offers the chance to jump forward, it assumes the branch will not be taken. If the predictions are correct, instructions are pre-fetched and loaded into the instruction queue and the correct data is available in the pipelines and no delay is experienced. If incorrect, the pipeline has to be flushed and reloaded losing several clock cycles.

In the case of unconditional jumps, the program just tells the microprocessor to move to another section of the program and no choice is involved. If the jump is to a distant address, the relevant instructions may not be in the cache and the cache would have to be flushed (re-loaded) (see Figure 13.2).

Figure 13.2

Branch prediction

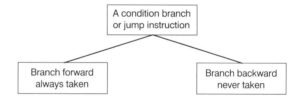

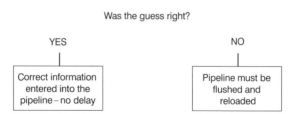

Integer unit and registers

As expected with a RISC processor, there are plenty of registers. In this section of the 601, we have 32 registers, each 32-bits wide. These registers are dual-ported. This means that two circuits can access the registers at the same time without interfering with each other. This is

187

like someone reading the back of your newspaper as you are reading the front – except that registers don't find it irritating. 'Port', by the way, is just a fancy electronic word meaning 'connection'. Transistors, generally, have three wires going to them and so are described as three-port devices.

The integer unit handles all instructions like integer arithmetic bit manipulation and transferring data to and from the external memory and is organized into a three-stage pipeline. In Figure 13.3, the second clock pulse executes the first instruction. The next clock pulse executes the second instruction and the last clock pulse executes the third. We have achieved the target of one clock pulse per clock pulse. And in the fourth clock pulse, we can see the next instruction just arriving to be decoded immediately after the first write-back.

Figure 13.3

Integer unit pipeline

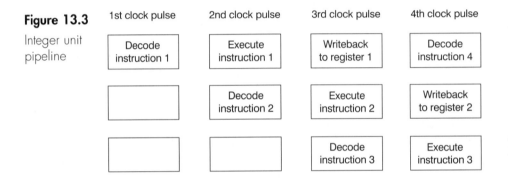

Floating-point unit

This has a further 32 registers but in this case, they are 64-bits wide and to fill a register with a single clock pulse, there is an internal 64-bit bus connecting it with the cache. The pipeline is five stage: prefetch, buffer, decode, execution and write-back.

Memory buffer

This acts as a buffer for the external memory. The buffers include two reads and three writes, each up to 32 bytes. It is also used in writing-back to the cache.

Big and little endians

The main memory is divided into locations each having its own address. Each location can hold a single byte of information. If we wanted to store a 32-bit number, then we would have to utilize four consecutive locations.

Imagine that we wished to store the 32-bit number 00000000 01010101 00010001 11111111_2 and we had addresses 24646603H, 24646602H, 24646601H and 24646600H available. Little-endian format would store the most significant byte in the highest memory address so, in our example, the data 00000000 would go into address 24646603H. This is used by Intel microprocessors. Big-endian, which Motorola uses, works the other way around. The most significant byte is put in the lowest memory address so, in our example, the data 00000000 would go into address 24646600H. These are shown in Figure 13.4. All the PowerPC microprocessors are switchable to enable little or big-endian to be used.

Figure 13.4

Big and little endians

Address 4 is the highest address

Byte 4 is the most significant byte of the number to be stored.

PowerPC 970

A large number of PowerPCs have continued to power the Apple-Mac and IBM desktops and, in addition, support both the UNIX and Linux operating systems.

The latest offering is the 970 with its 52 million transistors started life as a 1.8 GHz device and has now progressed to 2.0 GHz. This may appear slow but it has compensating attributes such as its 900 MHz bus as opposed to the 533 MHz bus of the Pentium 4.

It is a 64-bit micro so it handles data in 64-bit chunks but remains compatible with earlier 32-bit designs. It has two level 1 caches, one for instructions at 64 kB and a data cache of 32 kB, which are somewhat larger that the Intel product but both companies use a level 2 cache of 512 kB.

As memory size is continuing to increase with each design, the size of memory that can be directly accessed increases with the move to 64-bit processing. The Pentium 4 can access 40 GB of memory, which seems excessively large at the moment but there was a time when 4 MB was something to wonder at. The PowerPC 970 can handle memory of Star Trek proportions measured in terabytes (thousands of Gigs).

Table 13.1 Cache sizes

	L1 Instruction	L1 Data	L2 cache
PowerPC 970	64 kB	32 kB	512 kB
Pentium 4	It's a secret	8 kB	512 kB

For maximum microprocessor speed we need a high clock speed combined with the maximum use being made of every part of the microprocessor. The early 8-bit microprocessors would accept the first instruction and it would pass through the microprocessor being decoded, then acted upon, then having the results stored before it considered the next instruction. This meant that each bit of the micro was doing nothing for much of the time.

Modern micros load many instructions at the same time and split up the tasks so that as many as possible can be carried out at the same time to have the minimum time wastage.

As with the Pentium 4, the PPC970 makes use of level 1 caches that, as is now common, are split into an Instruction cache and a Data cache. There is also a level 2 cache and an external level 3 cache.

Loading the instructions

The instructions pour down from the Instruction cache at a maximum rate of eight per cycle, though five is a more likely overall figure. But this is still fast.

The PP970 uses a very long pipeline and can be handling up to 200 instructions simultaneously. The price of such a long pipeline is that we must be careful to ensure that it is filled with the most useful instructions and hence we need to back it up with very effective branch prediction techniques.

Branch prediction

To obtain the maximum possible speed, the PP970 has devoted a great deal of resources into its branch prediction. As the instructions are loaded, the branch prediction circuitry scans the incoming instruction looking for branch instructions. Every time we meet a branch instruction that offers a choice of outcome the branch will have to be accepted or rejected.

The 970 has two branch prediction methods. The first is very similar to that used in the Pentium 4 and, to over simplify the situation, it follows the same sort of reasoning as we often adopt in everyday life. If it usually happens, it is most likely to happen again. The 970 keeps a record of the previous 16384 branches in its BHT (Branch History Table) to see how often each choice was made and then this information is further sorted by a prediction program before it comes to a final decision.

The second method involves a similar sized table called a Global Predictor. This method also comes up with a final go/no go for the branch but it decides by generating an 11-bit vector that stores the actual execution path taken by the previous eleven fetch groups leading up to the branch.

So there are two independent mechanisms that make a decision as to whether the branch should be taken. If they disagree, we need a referee. This job is performed by a 'Selector Table' that stores the success rate for each of the two previous methods for each particular branch. It then makes the final decision – and it is said (by IBM) to be very successful, which it probably is.

Handling the instructions

Having combined the incoming instruction stream from the Instruction cache with the information from the Branch predict, the instruction are queued and passed to the Decode, Crack and Group Formation Unit.

At this stage, in order to keep the instruction handling speed at a maximum, this unit takes the instruction codes from the Instruction cache, decodes them and cracks them into their component parts called Internal Operations (IOPs). These very small but simple tasks are passed out to specialized units like the five blocks shown along the bottom of Figure 13.5.

Figure 13.5

The PowerPC 970

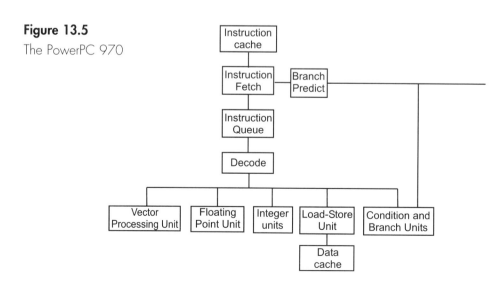

The IOPs are executed in whatever order that will result in the fastest throughput and to reduce the complexity of keeping track of the execution of each and every one, they are organized in groups of five and then the groups are tracked.

Of the final row shown, there are the arithmetically based block that handle the vectors, floating point and integer calculations, the load-store that handles the transfer of data to the memory via the second level cache and finally the feedback path for the branch prediction information.

The PC market place

The PowerPC may not be in our PC but it may well be in our car. The Ford Motor Company has elected to use the PowerPC as first choice for their engine management computer into the next century.

Quiz time 13

In each case, choose the best option.

1 The maximum number of instructions that the PowerPC 970 can be dealing simultaneously is:

(a) 200.
(b) 3.
(c) 16384.
(d) 128.

2 Write-back:

(a) reverses the order of the bits of data.
(b) is used to double-check the accuracy of data before use.
(c) is only used in the little-endian system.
(d) stores results in the cache rather than in the external memory.

3 The PowerPC 970 has an internal bus running at a frequency of:

(a) 64 bits/s although it can run at 32 bits/s.
(b) 512 kB/s.
(c) 900 MHz.
(d) 533 MHz.

4 A register that can be accessed by two circuits at the same time is referred to as:

(a) a second-level cache.
(b) dual-ported.

(c) a buffer.

(d) a three-ported device.

5   Big endian format:

(a) stores the low byte in the highest address.

(b) stores the high byte in the highest address.

(c) is used in all microprocessors.

(d) is used in a cache but never in the main memory.

14

The Athlon XP

This is AMD's competitor to the Pentium and is concentrating the mind of both companys and greatly benefiting the rest of us.

Competition concentrates the mind as well as improving things for the customers.

AMD has been creeping up on Intel for several years and finally the Athlon's 37 million transistors are giving the Pentium a serious problem. It is usually cheaper and, in many tests, faster. The thought behind the Athlon is not to compete in terms of clock speed but to go for real speed by doing more work for each clock cycle. Even so, the Athlon XP is now competing head-to-head on speed, having matched the Pentium at 2.8 GHz using the same 0.13 micron technology though with a different internal design and ensuring (of course) that the two microprocessors are not pin-for-pin compatible. The Athlon includes a similar system of protection against thermal overload as in the Pentium.

An outline of the Athlon XP is shown in Figure 14.1.

Caches

For maximum speed the caches are on-chip. This eliminates the travel-time delay as the data is moved.

From the external memory and the surrounding hardware, the incoming information from the system bus is fed into a 64 kB

Figure 14.1

The Athlon XP
processor

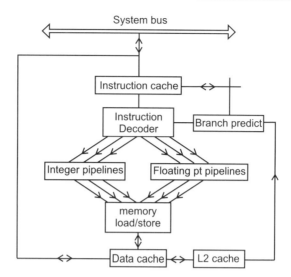

instruction cache and a separate 64 kB data cache. The data cache feeds data into the L2 cache, which is somewhat larger at 256 kB and has techniques to ensure that the L2 cache does not duplicate any of the information stored in the data cache and hence we effectively have a 384 kB local high speed storage area.

Branch predict

As with all current microprocessors, great care is taken to guess the likely result in each branch instruction. Such instructions produce, usually, two alternative routes for the program. They answer questions like 'is the result zero?' and the answer will determine what happens next. If we always wait until the question is answered and only then do we load the instructions for the next bit of the program there is much wasted time as we saw in the earlier microprocessor designs. If we guess correctly, we can already pre-load the next part of the program and get started on it. The branch prediction circuitry does the guessing. If it gets it wrong, the old data is ditched and replaced.

Hardware data prefetch

This is a further form of prediction similar in which the incoming instructions are monitored and, as they are still arriving, the data that will be needed is guessed at, and loaded into the data cache so the Athlon loads data before it knows that it will be needed. As with the branch prediction, incorrect data has to be overwritten but on balance, it speeds up the data flow.

Instruction decoders

To make full use of its slower clock speed, the Athlon has three instruction decoders that can run independently. Each of these can handle three operations per clock cycle giving an overall throughput of nine operations per clock cycle, which is still significantly greater than the six operations per clock cycle of the Pentium.

Pipelines and instructions

The Athlon has three independent integer pipelines and also three similar floating-point pipelines whereas the Pentium has four pipelines for integers but only two for floating points.

The three floating-point execution units simultaneously handle:

(a) store and load functions
(b) add functions
(c) multiply functions such as all the Intel MMX (multimedia extensions) instructions plus AMCs own SIMD (single instruction multiple data) instructions to provide full support SSE (streaming SIMD extension) and more lifelike 3D imaging and graphics – AMD's name for these new instructions is '3D NOW!' technology. (MMX is an Intel trademark; 3D NOW! is an AMD trademark.)

The state of the competition

The Pentium had a 'rapid execution engine' which had two ALUs (arithmetic and logic units) for the integer instructions, each clocked at twice the core processor speed running a front side bus at 533 MHz whereas the Athlon XP had only a 333 MHz FSB. This continues the pattern of the Pentium claiming the headline figure for speed. However, on balance, the Athlon is, by most tests, slightly faster than the Pentium.

An update . . .

That was written yesterday. This morning came the news that Intel has burst through the 3 GHz barrier (just) with a 3.06 GHz device. This, they say, includes hyper-threading, a technique that involves splitting a program into units that can be ran simultaneously. It allows the micro to run multiple applications at the same time, with the processor appearing to be two processors. Such multitasking is available in Windows XP and Linux and probably all their successors. So where does this leave the future, are we going to go for greater and greater speeds, or will we develop multi-tasking so we effectively have greater and greater numbers of micros sharing the work? I have a feeling that task sharing will be the answer.

It seems likely that Intel is now back out in front.

Exciting times ahead . . .

Another update . . .

Almost immediately, Athlon has replied with what appears to be another significant step forward – 64-bit computing.

The microprocessor which as yet has been living with the codename 'Hammer' will be sold as the more user-friendly name of 'AMD Athlon 64' and will be available in mid-2003 and will join the PowerPC 970 in the '64' club. It will be able to run 64-bit, 32-bit and 16-bit applications without any speed penalty and so avoid the cost of buying new software.

The only technical information that is included in the initial announcement is a new bus system using 'hypertransport' technology which AMD claims to increase throughput by 50% over existing designs. Intel will have something to say about that claim, I expect. The clock speed of the first batch will be little different from the XP, around the 2.8 GHz, but the design will provide more scope for development and will be able to run programs at a higher speed.

Really exciting times ahead . . . over 3 GHz clock speeds, 64-bit computing and multiple instructions being carried out simultaneously. Sounds good.

The desktop speed Olympics is shared between the PowerPC 970, Pentium 4 and the Athlon 64 whereas the computer market is dominated by the IBM clones leaving just a minor role for the PowerPC 970 in the Apple-Mac. As we saw earlier, the result of any speed test does depend on the nature of the test. Having said that, and at the risk of irritating the fans of each, in the race for the overall speed freak the Athlon 64, when it is available, will appear to be the winner with the other two virtually shoulder to shoulder a pace behind. But it depends on the test chosen and we know that any speed king will be dethroned so very quickly.

A (very) approximate comparison based on the currently available information is shown in Table 14.1.

Table 14.1

	PowerPC 970	Pentium 4	Athlon 64
Clock speed	2 GHz	2.8 GHz	2 GHz
Bus speed	900 MHz	533 MHz	533 MHz
Bits	64	32	64
Process size	0.13/ 0.09 microns	0.13 microns	0.13 microns
Op systems	OSX IBM linux	Windows	Windows
Comparative speed	1988	1984	2372
Max memory	Terabytes	40 GB	Terabytes

Quiz time 14

In each case, choose the best option.

1 Compared with the Pentium 4, the Athlon XP design has:

(a) faster FSB, running at 533 MHz.
(b) the same speed of FSB.
(c) slower FSB, running at 333 MHz.
(d) faster FSB, running at 2.8 GHz.

2 As the Pentium 4 and the Athlon XP are both using 0.13 micron technology:

(a) it does NOT imply any other similarities between the designs.
(b) they will both run at the same clock speed.
(c) they will have the same number of pins.
(d) the cache sizes are equal.

3 The three floating point execution units in the Athlon XP simultaneously handle store and load, multiply functions and:

(a) SIMD functions.
(b) add functions.
(c) divide functions.
(d) 3D NOW! functions.

4 When Branch prediction is correct it:

(a) increases the overall speed of running the program.
(b) increases the length of the pipeline.
(c) decreases the clock speed.
(d) prevents overheating of the microprocessor.

5 The Athlon Instruction cache has a capacity of:

(a) 256 kB.
(b) 32 bits.
(c) 64 kB.
(d) 384 MB.

15

Microcontrollers and how to use them

Getting ready for takeoff

In the 1960s, electronics started to awake from its slumber that had used thermionic valve technology that was recognizably similar to circuits that had been built for thirty years. The pace of progress was gentle. The first semiconductor material was developed and the transistor came into use in just a few years. The photographic process used to design and produce the transistor quickly led to simple integrated circuits and the microprocessor.

The start of the microcontroller

No sooner had the microprocessor and the associated memories arrived in 1971 than it became obvious that the microprocessor was always accompanied by other circuits, like input/output devices, memory and timing circuits so it would be a good move to combine them into a single device.

We had a choice – we could keep everything general and universal and call it a microprocessor or design it for a single purpose and call it a microcontroller.

The multipurpose devices went into computers and even here we had a choice. Computers were either 'microcomputers' where price was a significant feature and these microprocessors had some built-in ROM and RAM. Soon, however, speed became the main feature as the prices began to fall and we could afford to equip our homes with computing power which equalled many offices of just a few years previously and the microprocessors became expensive and fast. Speed headlines drive the publicity machines as home and office computers became faster and faster. They sold in their millions.

Meanwhile the single-purpose devices, really the descendents of the early microcomputers, were developed further and made really cheaply, sold by the billion and were never mentioned. They power the pocket calculator, video recorders, cameras, microwaves, washing machines and greetings cards that play music – in fact almost anything vaguely electronic.

Just a thought

The microcontrollers outnumber the population of the world many times and as mentioned earlier we are likely to be sharing our homes with, possibly, fifty of them. They are in every essential industry – food production, transport, communications, research, weaponry, power generation, medicine, heating and air conditioning – there is little that we rely on that does not use a microcontroller. If they learn to communicate independently of us, they may develop their own agenda. Now there's a thought.

Most microcontrollers are similar

Once we have learned to drive, most vehicles are easily recognized as being very similar. We are happy with the general idea and can concentrate on the minor differences. Microcontrollers are much the same. Having already become familiar with the basic building blocks of the simple microprocessor in Chapter 8, we can move very easily into the microcontroller. It is not surprising then to find that all microcontrollers are basically very similar.

To give an overall impression of the range of microcontrollers available we are going to look at three popular ranges. The first is the 8051, probably the most widely used microcontroller, over twenty years old and continuously developed by many different companies and showing no signs of fading away. The next is from the AVR family produced by the Atmel Corporation, one of the leaders in this field. From this range we look at the AT90S/LS2343 one that is small, modern and RISC. The final one will be explored in Chapter 16.

The 8051

Probably the transition between the microcomputer to the micro-controller occurred with the Intel 8048 as we saw in Chapter 11. The 8048 added on-chip RAM, ROM and a timer so it could be used as a single purpose device such as controlling a keyboard – it was, in fact, a microcontroller.

With the experience gained by using this, it became apparent that there was a significant market for a microcontroller.

In 1980, Intel launched the 8051 which, twenty-three years later, is alive and well. In fact very well indeed. It is probably the most popular microcontroller ever. It is made by about 44 suppliers.

These suppliers have often added some extra features to make versions or 'variants' as they are called particularly suitable for specific jobs. There are at least 92 variants all compatible with the original code. Even within variants, there are a series of options that lifts the total number of members of the 8051 family to several hundred.

Numbering

The device numbering is not very obvious as many microcontrollers are available from several different suppliers with their own product code. They then produce a group of basically similar devices with minor changes like different operating voltages or differing amount of RAM and ROM on-board memory – these groups are referred to as 'families'.

The family is given a name which often has little connection with the product codes. For example, Intel's 8051 family has the family name of MCS51. This contains the 803X, 805X, 875X and the low power versions bXC45X and the 8XCX52. As usual the X refers to any figure or letter in that position.

The situation is further confused (or possibly simplified) by referring to all of them as 'the' 8051.

The block diagram of the 8051

The block diagram shown in Figure 15.1 is the family portrait of the 8051 family. There are some features that differ between the family

Figure 15.1
8051 block diagram

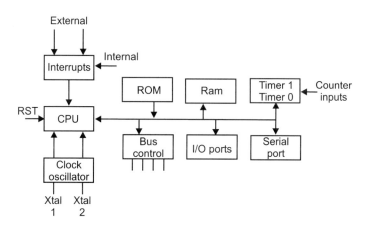

members, principally the memory configuration – some versions have less memory and some have none at all. However, we will look at the operation of our 'middle-of-the-road' version and worry about the individual differences later.

The 8051 pinout

As in all micro and digital chips, a line over a pin designation indicates that it is active low or, put more simply, to use this feature we need to apply zero volts.

The pinout shown in Figure 15.2 looks, at first glance to be rather complicated due to the dual use of many pins. This is a common feature of microcontrollers as a method of reducing the number of pins to be used. The more pins, the more expensive and the larger the device. This is bad news for a device often destined to be embedded within another circuit.

Figure 15.2

Pinout of the 8051

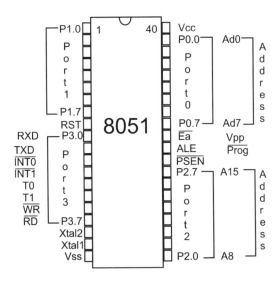

There are variants available that provide the increased number of pins so that there is a separate pin for each function.

Reset

Regardless of what program is being run, we must always be able to gain control of the microcontroller just as we must with a micro-processor. The procedure is just the same. The microcontroller has a reset pin which, in the 8051, is taken from the bus control block and, in normal operation must be held to zero volts. When a positive voltage over 2.0 V is applied to it the microcontroller immediately returns to its startup memory location, which in this case is 0000H. We can arrange this to occur automatically when the power is switched on but we should also provide a reset switch to gain control of the system at any time without removing the power. This is the 'reset' switch which we use when our computer locks up and ignores us.

When changing microcontrollers, remember to check the polarity of the reset voltage. Compare this circuit in Figure 15.3 with the one shown in Figure 8.7.

Figure 15.3

The reset switch

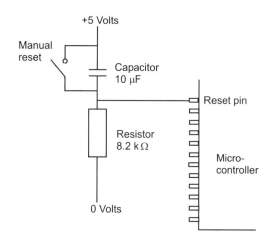

Clock input

As we have seen in Chapter 7, we are going to need a clock signal. The original design of the 8051 called for a 12 MHz crystal though later versions can run at 33 MHz, 40 MHz or even 44 MHz. As an alternative, we can use a ceramic resonator or an external signal. The clock input is shown in Figure 15.4 using a crystal. To use an external signal throw away the crystal and the capacitors then apply the external signal to the Xtal1 pin and leave Xtal2 disconnected.

Figure 15.4

The clock

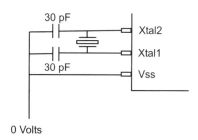

Ports

We have four ports numbered 0 to 3. In the standard 8051 ports 0, 2 and 3 are dual purpose and port 1 is just an input/output port. In some variants, all ports are dual purpose.

With regard to the memory, we can use the on-chip memory, which keeps the circuit as simple as possible and makes the embedding of the device at its most convenient.

The on-chip memory may be masked ROM or an EPROM which we met in Chapter 6. To use external memory obviously increases the chip count but can allow up to 64 kB of memory, which is the standard size for 8-bit microprocessors.

The external memory is accessed by taking the 'external access' (EA) pin to a logic zero. To switch the external ROM on, the output enable (OE) pin of the external memory is taken low by the program store enable (PSEN) pin of the 8051. In a similar way, an external RAM is accessed for reading and writing by the read (RD) and write (WR) pins applying a logic zero voltage to the output enable (OE) and read/write (R/W) pins respectively.

Interrupts

The 8051 has a total of five interrupt signals. Two of these can be externally generated and three are of internal origin. Interrupts are discussed more generally in the chapter dealing with interfacing but basically, when an interrupt occurs, the program which is running at the time is interrupted and another code sequence is ran, called an interrupt service routine (ISR). When the ISR is compete, the microcontroller returns to its original program and continues as if the interruption has not occurred. There is a different ISR for each interrupt which must be pre-loaded in specific memory addresses, all ready to go if needed.

What if two interrupts occur at the same time?

The interrupts are checked continuously in what we call the polling order. Starting from the top, the order in the 8051 is:

External Interrupt 0
Timer 0
External Interrupt 1
Timer 1
Serial port

In addition, each interrupt can be given two levels of priority so if two interrupts occur, the one with the high priority will be handled first. If two have the same priority, it is decided by the polling order.

Timer/counters

A timer or counter is a series of bistables or flip-flops that change state once for every input signal, thus one of these circuits would divide the input frequency by a factor of two. If this signal is then fed into the next bistable, the output is $\frac{1}{4}$ of the original frequency. The next circuit would have an output of 1/8, then 1/16 and so on.

There are two timers, T0 and T1. These can be programmed to divide by 256, 8192 or 65536 and will generate an interrupt signal upon completion that can be detected by the software. One of the modes allows the timer/counter in 8-bit (divide by 256) mode to reload and start counting again each time continuously.

The input signal being counted can originate from an external circuit so it counts the number of incoming pulses, or it can use an internal signal which is actually 1/12 of the clock frequency in use. As mentioned above, it generates an interrupt signal when it reaches its maximum value. We can preload the timer with a number to start counting from. This will allow the interrupt to be generated after any required number of events, or time interval.

Serial port

Since the microcontroller normally handles data eight bits at a time, it is operating in parallel but two receive or transmit serial data we have perform serial/parallel conversions. This is always achieved by using a shift register working under the control of a clock signal. The working of a shift register is described in Chapter 17.

The serial port is able to transmit and receive data, at the same time, this is referred to as a full duplex system. As the name suggests, the bits of data are moved one after the other in a continuous stream. Pin 10 is called RXD or receive data and pin 11 is the TXD, transmit data but, as is often the case, things are not quite that simple.

It can operate in three ways, or modes numbered 0, 1, 2 and 3.

Mode 0

This is the case that spoils the simple RXD, TXD as, in this mode only, the TXD pin is actually used as a clock signal and the RXD is used to receive or transmit data. The clock frequency is fixed at 1/12 of the onboard oscillator frequency and this, of course, determines the speed at which the data is transferred via an 8-bit shift register.

Mode 1

This mode also sends data in 8-bit lumps but its frequency is adjustable and operates as an 8-bit UART (universal asynchronous receiver transmitter – see more about this in Chapter 17). The 8 bits are increased to 10 bits buy adding a logic 0 to indicate the start of the group and a logic 1 to mark the end of the group. This is the normal format used in RS232 transmissions. Unfortunately, the output voltages do not comply with the RS232 standards so an external chip must be added to do the voltage conversion. Some suitable chips are discussed in Chapter 17.

Mode 2

This is very similar to Mode 1 except the number of bits transmitted is increased from 10 bits to 11. The extra bit can be used as a parity bit which is used to check for transmission errors, The pattern is Start bit (0), 8 data bits, parity bit and stop bit (1). (Have a look at Chapter 17 again.) The transmission rate can be 1/32 or 1/64 of the onboard oscillator frequency.

Mode 3

This is an 11-bit transmission with a programmable baud rate. The baud rate is near enough the same as the transmission rate measured in bits per second.

Watchdog timer

When a microcontroller is embedded in equipment it may find itself used in areas where electrical interference is a problem. This, or a software problem can cause the microcontroller to lock up by getting into an endless loop. The watchdog timer will reset the micro-controller after a period of time, about 20 milliseconds, unless it is told not to. Stuck in a software loop, it no longer generates this 'don't reset me' signal and so the microcontroller is reset and escapes from the loop.

The watchdog timer is only fitted into some of the 8051 variants but is available as a stand-alone chip but is commonplace in newer designs.

When we want to leave the microcontroller in a continuous loop, we have a choice of ensuring that the loop contains the required software code or disabling the watchdog timer before the loop is started.

AVR 8-bit RISC microcontrollers

Taking the AT90S/LS2343 as an example, we can see a really basic microcontroller with minimal complexity yet having many useful features that make it inexpensive, small and comparatively fast. The RISC design and the width of the registers allows the vast majority of instructions to be executed in a single clock cycle and all the others, apart from five, are completed in two cycles.

Figure 15.5

AVRAT90S/LS2343

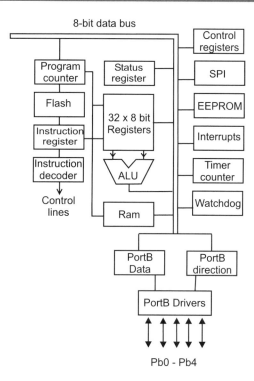

Pb0 - Pb4

In Figure 15.5, we have a block diagram. It is quite a relief to see that mostly it is quite familiar after looking at the earlier 8-bit micro-processors and the 8051 considered in previous chapters. Already we see that most devices are a combination of one or two innovative ideas added to a standard mix.

Inputs and outputs

One notable feature of the block diagram, we see only five input/outputs shown as PortB 0 to 4 so we have only five connections for data which is a small number until we look at the sister version, the AT90S/LS2323 which has only three – PortB 0 to 2! Yet they are called 8-bit devices and in previous 8-bit micros we have grown to expect at least one and sometimes two 8-bit input/output connections. The answer is just that the '8-bit' description refers to the internal data bus width.

The PortB pins can be used to send data in either direction, they can be used as inputs or outputs. As is common with other microprocessors and microcontrollers, the direction of data movement through each of the PortB lines is individually controlled by a data-direction register. Loading a 'one' into the data-direction register will make the corresponding PortB line into an output, on the other hand a 'zero', of course, will change it into an input.

Memories

This microcontroller has three memories, or four if we count the general purpose registers. The loading of the memory storage areas employs a serial transfer of data and is achieved by the SPI (serial programming interface).

One ROM storage area is achieved by a Flash memory organized as 1k × 16. The program instructions are all either 16 or 32 bits wide and can be cleared and reprogrammed whilst remaining in circuit. It can be cleared and reprogrammed at least 1000 times. The contents of the Flash memory cannot be changed by the program being executed by the CPU and so is free from accidental corruption.

The data is held in an EEPROM, which again can be cleared and reprogrammed electronically without removal from the device. It only holds 128 bytes of memory but is able to go through at least 100 000 cycles.

Data corruption can occur in the EEPROM if the supply voltage is reduced too far but this effect can be avoided by any one of the following three methods. The first we have already mentioned – use the Flash memory for critical data. The other two methods are ways to detect the reduction in voltage and immediately put the micro-controller into a safe condition. This is often referred to as 'brown out protection'. An external circuit detects the falling voltage and applies a low voltage to the reset pin which effectively switches the chip off until the supply voltage recovers. The alternative is to put the microcontroller into a power-down sleep mode which is a power saving mode which has the effect of preventing any decoding or execution of any instructions – which, of course, precludes any 'writes' to the EEPROM.

It also has 128 bytes of SRAM (Static RAM) for the temporary storage of data and 32 8-bit general purpose registers that can be connected two at a time with the ALU (arithmetic and logic unit) which is the heart of the 'brain' within the microcontroller.

Clock

The AT90S/LS2343 has an internal RC oscillator which runs at 1 MHz, 4 MHz or 10 MHz depending on the version in use. It is one of the few of the micro devices that does not make use of an external crystal although it can use an external clock pulse. This external clock pulse only requires a single pin and hence we have an extra pin to use as an output.

Interrupts

There are only three interrupts. The first, and highest priority, is the reset which is activated by a low voltage applied to pin 1, a power-on reset or a signal from the watchdog. The next is an external interrupt request as described in a moment. Lastly, an overflow from the timer/counter circuit.

Pinout and package

These are shown in Figure 15.6 and we can see that it is available as an 8-pin DIL (dual in line) package which has two lines of pins and also the surface mount version, plastic gull wing SOIC (small outline IC).

Figure 15.6

AT90S/LS2343 pinout

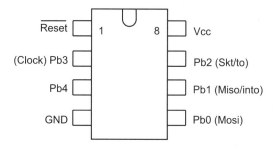

Pin 1 – Active low reset. Must go low for at least 50 ns.
Pin 2 – External clock signal input or PortB 3
Pin 3 – PortB 4. All lines can sink 20 mA and therefore are able to power LEDs directly. Sinking means that the LED or other load is connected between the positive Vcc supply and a low voltage output at the port.
Pin 4 – Ground.
Pin 5 – PortB 0 or MOSI. In serial programming mode, MOSI is the serial data input.
Pin 6 – PortB 1 or MISO/INT0. In serial programming mode, MISO is the serial data output. This pin can also act as the external interrupt described in the previous paragraph.
Pin 7 – PortB 2 or SCK/T0. In serial programming mode, SCK is the serial clock input. This pin can also provide the timer/counter0 clock input.
Pin 8 – Vcc. The LS version requires a positive supply voltage that remains in the range 2.7–6.0 V.

Sleep modes

When the microcontroller is not being used, it can switch off some of its circuitry to save power. Sleep modes are employed in all modern microcontrollers and make an enormous difference to the overall life of an intermittently used device. This enables sealed units in toys and greetings cards to remain active for months or years.

Idle-mode

To see the benefits, this microcontroller has a normal operating current drain of 2.4 mA but when switched to the 'idle' mode, the current falls to 0.5 mA. It does this by stopping the CPU activity but allows the timer/counter, watchdog and interrupts to remain operational. This is about an 80% power reduction but we can do a lot better than that otherwise my musical socks would have stopped long ago.

Power-down mode

In this mode, only the external interrupt and watchdog (if enabled) continue to work and current falls to less than 1 microamp, which is a really impressive reduction in power. The microcontroller can be aroused from its sleep only by one of the following: an external reset, the watchdog (if enabled) or INTO external interrupt.

The PIC16F84A

This is another modern RISC development and has many features that are similar to the AVR that we have just looked at. The AT90S/LS2343 was chosen as representative of the very small and basic micro-controllers found embedded in many products. This PIC16F84A example from Microchip Technology is a mid-range device which is larger and more capable than the AVR.

The PIC series ranges from a really simple 8-pin, 4 MHz micro-controller on a level with the AT90S/LS2343 that we have just considered up to a 40-pin 25 MHz device. As mentioned the PIC16F84A is a mid-size version that has 18 pins and runs up to 20 MHz.

The PICs are designed for easy use and are becoming increasingly popular as the first step into the world of microcontrollers. Microchip Technology provides a PICSTART™ PIC development system that provides, at a very reasonable price, an assembler, compiler, EPROM and EEPROM programmer, all the hardware manuals and even a sample PIC to play with. It should be mentioned that other companies have similar systems compatible with the PIC series and for other microcontrollers like the AVR and 8051 series.

It has proved to be such an easy, off the shelf, starting point that to many people 'PIC' is not only their first choice but is becoming used as a generic term for any microcontroller.

The general layout of the PIC16F84A is shown in Figure 15.7.

Figure 15.7

PIC16F84A

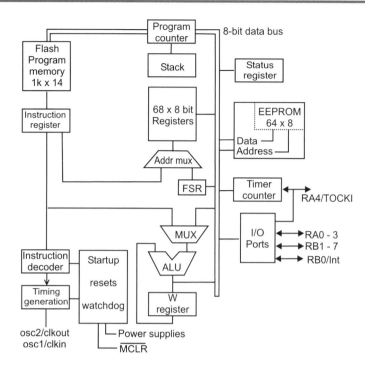

Supply voltage

The DC supply voltage must remain in the range 2.0–5.5 V for it to work happily though similar versions such as the PIC16F84 can run between 2.0 and 6.0 V.

Sleep mode

To save the current drain, a software instruction can put the PIC into a sleep mode. The supply current is very dependent on the clock frequency and is normally between 1 and 20 mA and when put to sleep the drain is reduced to approximately 1 µA.

To obtain the lowest possible sleep current we should hold all I/O pins at V_{DD} or V_{SS} and disable any external clock and hold the master clear pin in a logic high state – not in the reset state. The purpose of all this is to prevent any voltages from floating up and down. If it did so, it would switch and the technology used results in very low currents drawn except at the moment of switching during which it causes a really high spike of current so the higher the frequency, the more often this spike occurs – hence the increased current.

The microprocessor will wake if: a reset occurs by a logic low voltage being applied to the MCLR pin, a wakeup pulse arrives from the watchdog unit (if it is enabled) or an 'EEPROM write complete' signal.

211

Memory

As usual, we have two blocks of memory. One is the program memory and the other is for data. For maximum speed, they each have a separate bus connection so that both memories can be accessed during a single clock cycle.

Program memory

The program memory is situated in the flash memory which is organized as 1028×14. All instructions in the PIC16 series use 14-bit instructions.

The reset vector points to address 0000H and the interrupt vector is 0004H so address locations 0005H to 03FFH are available for us to hold our programs.

Data memory

The data area is subdivided into two areas, the FSR (file select register) and the GP (general purpose registers) as shown in Figure 15.8.

Figure 15.8

Register file map

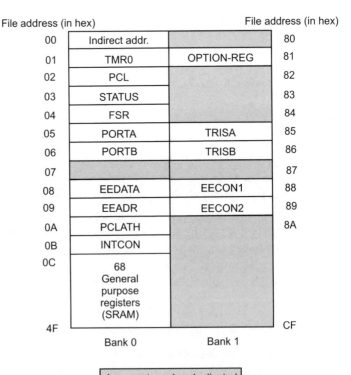

File address (in hex)

Address	Bank 0	Bank 1	Address
00	Indirect addr.		80
01	TMR0	OPTION-REG	81
02	PCL		82
03	STATUS		83
04	FSR		84
05	PORTA	TRISA	85
06	PORTB	TRISB	86
07			87
08	EEDATA	EECON1	88
09	EEADR	EECON2	89
0A	PCLATH		8A
0B	INTCON		
0C	68 General purpose registers (SRAM)		
4F			CF

Bank 0   Bank 1

Areas not used or duplicated

The SFR (special function register)

This register controls the operation of the CPU and involves such things as the input and output ports, EEPROM address and data, timer, program counter and that sort of housekeeping.

All the register files are 8-bits wide and are arranged in two banks called bank 0 and bank 1. We have to instruct the microcontroller as to which bank is to be used and this is done by using special instructions to access some of the page 1 registers. Those accessible are indicated in Figure 15.8. Microchip Technology are planning to remove the choice of using the OPTION and the two TRIS registers and suggest that the STATUS register is used instead. This does not affect the use with this chip but it will ensure that upgrading in the future will not require any modifications to the software.

I/O (input/output) ports

All outputs can source or sink 25 mA and can therefore power significant external circuits without further power amplifiers being required. Sinking means that the load is connected between the positive Vcc supply and a low voltage output at the port and sourcing is connecting the load between a positive output on the pin to the ground.

PortA and TRISA registers

PortA is a 5-bit wide bi-directional port, each line being individually controlled so some of the lines can be inputs whilst the others are outputs. The choice of input or output is made by loading a 0 (output) or a 1 (input) into the appropriate bit of the data direction register TRISA.

In common with other devices, when it first starts at power-on, the port is set as an input. This provides a safer option that running the risk of random information being sent out to whatever it is connected to.

PortB and TRISB registers

PortB is a 8-bit wide bi-directional port, each line being individually controlled using TRISB in the same manner as in PortA. Each of the PortB pins have a weak internal pull-up which can be switched on or off by the RBPU of the option register. The pull-ups are disabled when the port is being used as an output and also during power switch-on.

Any of the Pins RB4–RB7 that just happen to be configured as an input have an interrupt-on-change feature that can be useful. If any one or more of these pins have changed logic state since they were last read, it causes an RB port change interrupt. This interrupt can be used to wake the microcontroller from its 'sleep' mode.

Status register

This is very similar to the one we met when looking at the Z80 in Chapter 8.

The bits are:

Bit 0 is the C(carry/borrow) bit. 1 = a carry out from the MSB (most significant bit) of the result otherwise it is cleared to zero. For 'borrow' the values are reversed. Subtraction is carried out using the two's complement method that we met in Chapter 4.

Bit 1 reflects the carry situation that last occurred from the 4th bit of the result. This is also called the half-carry bit.

Bit 2 is the zero flag. It is set to ONE when the result of the last arithmetic or logic operation is ZERO. Be careful not to misread this.

Bit 3 goes to 0 after running the SLEEP instruction.

Bit 4 goes to 0 when a watchdog time-out has occurred.

Bit 5 is used to select between the two memory banks. It is cleared to 0 to access Bank 0 and set to 1 if we need access to Bank 1.

Bit 6 and bit 7 – not used. It will be used in the future so by programming them for 0, future compatibility will be assured. This may save a lot of time if our program is used on an upgraded version.

Option register

As the name suggests, it offers a series of options. One example is the control of the prescaler.

The Prescaler

Two functions are affected by the prescaler, they are the timer, TMR0 (timer zero) and the watchdog timer. Each of these circuits provides an output pulse after the count overflows and restarts from zero. In the case of the watchdog, the time interval is about 18 ms. If a longer time interval is needed, we have three alternatives. We can simply switch the watchdog off but, of course, we lose the benefits of the watchdog if the microcontroller gets caught in a loop. A simple way is to place the software code CLRWDT (clear watchdog timer) in the program at anytime before the end of the countdown so it is reset for another 18 ms and we can repeat this as necessary. Lastly, we can use the prescaler to reduce the frequency of the incoming pulses and hence increase the time before the output signal is generated.

Bits 0, 1 and 2 PS2:PS0 (prescaler rate select bits) provide eight alternative pulse rates for use by the watchdog or TMR0. Setting the three inputs to 000 will provide a clock signal which is equal to an external signal or chip oscillator. Changing the setting to 001 will halve the frequency (or double the time). Increasing the setting to 010 will decrease the frequency to one quarter of the original. As the count increases, the frequency will halve for each count until we reach the maximum bit value of 111 which will cause the watchdog time to increase by a factor of 128. As it happens, the TMR0 has a divide by two circuit built in all the time so a setting of 000 will halve the frequency and the maximum count will reduce the frequency by a factor of 256.

Bit 3 PSA – (prescaler assignment bit) – Unlike most other PICs, the prescaler can be applied to the TMR0 or the watchdog, but NOT both, so the option register controls the choice. Bit 3 of the option register is set to 0 to prescale the TMR0 and a 1 selects the watchdog.

The other bits

Bit 4 T0SE (TMR0 source edge select). This controls the moment of timing the clock input. A '0' increments the count on the low-to-high transition and a '1' increments on the high-to-low transition.

Bit 5 T0CS (TMR0 clock source select). This decides where the clock pulses come from. The choices are a '0' to an internal clock as on the CLKOUT pin and a '1' counts the transitions on the RA4/T0CKI (timer zero clock input). This option allows pulses to be generated by any external source like cans of beans moving along the conveyor belt or the revolutions of an engine.

Bit 6 INTEDG (interrupt edge select). This is similar to bit 4 except we are controlling the moment at which an interrupt signal is recognized of the RB0/INT pin. A '0' uses the falling voltage edge and a '1' sets the rising edge.

Bit 7 RBPU (PortB pull-up enable). This controls the 'pull-ups' on the PORTB output. A '0' allows each line to have its own pull-up enabled or switched off as required. A '1' switches them all off. A pull-up circuit is shown in Figure 15.9. When the switch is closed by applying a '0' state to this pin it connects the output to the positive supply via a current limiting resistor. This ensures that in the absence

Figure 15.9

A pull-up circuit

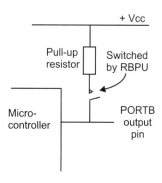

of any input data the port will be pulled up to a positive state so that it doesn't wander about applying unpredictable inputs. The higher the value of the resistance used, the easier an incoming voltage finds it to bring the voltage down and this is referred to as a 'weak' pull-up like this one. Likewise, a reduction in the resistance value will make the port input more determined to stay high and this is referred to as a 'strong' pull-up.

PCL (program counter low)

PC is the program counter that keeps track of the instruction being executed at the time. It is a 13-bit register divided into PCL for low byte and PCH for high byte. Its behaviour is common with other microcontrollers and microprocessors as described in Chapter 8.

Stack

This register stores the return address when interrupts occur. It stores up to eight 13-bit addresses. Again, there are more details in Chapter 8.

Register 6-1: PIC16F84A configuration word

This is a special register which is just memory location 2007H which can only be accessed during programming. Unprogrammed pins are read as '1'.

It is a 13-bit word with the following options:

Bit 13–4. This protects the program code from being read after the programming is complete. 1 = no protection 0 = all program memory is code protected.

Bit 3 Power-up timer enabling. This allows a nominal 72 ms delay when first switching the microcontroller on to allow power supplies to settle. The delay can be activated by loading bit 3 with '0', no delay = '1'.

Bit 2 watchdog timer, disable with '0', enable with '1'.

Bits 1–0 oscillator selection bits
00 = LP oscillator
01 = HS oscillator
10 = XT oscillator
11 = RC oscillator

Oscillators

The crystal or ceramic resonator options are shown in Figure 15.10 and the crystal oscillators are divided into the frequency ranges:

LP low power crystals 32 kHz–200 kHz.
XT crystal/ceramic resonator 100 kHz–4 MHz.
HS high speed crystal/ceramic resonator 4 MHz–20 MHz.
RC resistor/capacitor.

Figure 15.10

Oscillator options

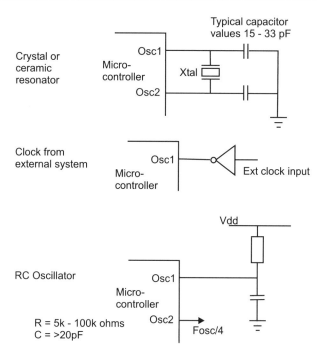

A note on ceramic resonators

These offer an alternative to crystals for oscillators. They are generally lighter, more shock resistant and are usually somewhat smaller and cheaper to buy. The downside is that they do not have quite the frequency accuracy or stability and in many respects they are positioned somewhere between an RC network and the quartz crystal.

An interrupt summary

The PIC16F84A has four sources of interrupts:

(i)   External interrupt on the RB0/INT pin.
(ii) TMR0 overflow.
(iii) PortB change detector pins RB7–RB4
(iv) EEPROM data write complete interrupt. This is used only when we are loading new programs into the EEPROM.

Quiz time 15

In each case, choose the best option.

1   An 8-bit microcontroller has:

(a) an 8-bit data bus.

(b) eight ports.

(c) an 8-bit address bus.

(d) eight internal registers.

2 Compared with the Pentium 4, a microcontroller:

(a) is faster.

(b) consumes less current.

(c) is more complex.

(d) is larger.

3 A watchdog circuit prevents:

(a) damage due to excessively high supply voltages.

(b) an overrun in timer/counter circuits.

(c) burglars.

(d) wastage of power while the microcontroller is not being used.

4 It is NOT essential for a microcontroller circuit to include:

(a) registers.

(b) a reset pin.

(c) some memory.

(d) a crystal.

5 The term 'ISR' refers to an:

(a) immediate service register.

(b) internal system reset.

(c) interrupt service routine.

(d) instruction set register.

16

Using a PIC microcontroller for a real project

PIC microcontrollers are a very convenient choice to get started in this field although, as we have seen, there are enormous areas of overlap between microprocessors and other microcontroller designs.

One convenience is that Microchip Technology has taken the RISC concept seriously. There are only 35 instructions and, of these, only a few are required to write quite usable programs.

All data movements are based around just a single register called 'W' for 'working'. This performs much the same function as the accumulator in the earlier processors like the 6502.

Getting started

To program the PIC we need a copy of software instructions and more details of the register layout.

There is no getting away from the fact that when we first meet a microcontroller, the information appears overwhelming – not only in the quantity but in the apparent complexity. This is despite the efforts of the designers to make it as simple as possible. It is very much like seeing a new foreign language, which it is really – and tackle it in much the same way, a bit at a time.

There are four golden rules which may help:

1 Don't get scared! From time to time, it will seem tempting just to throw it all away and find something easier to do.
2 Start really simple – one step at a time.
3 Be prepared for it to take a long time in the early stages.
4 When you are feeling low and despondent, remember that everyone else has felt the same. It will get better. Honest.

The hardware

Let's keep it really simple. We have to buy a microcontroller and connect it up and then program it.

The device that we are going to use is a PIC16F84A-04/P. The pinout is shown in Figure 16.1. The number PIC16F84A is the type of microcontroller. The 04 tells us that the maximum frequency is 4 MHz and the /P means that it is the standard plastic package. There is no minimum operating frequency and the slower it runs, the less power it consumes.

Figure 16.2 shows the most basic circuit. It includes only the chip supplies and a positive supply to prevent the MCLR (master clear) from

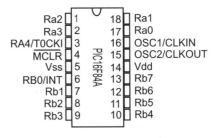

Figure 16.1

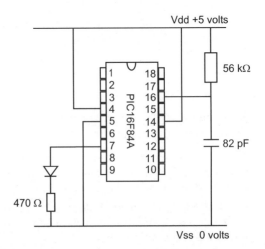

Figure 16.2

accidentally resetting the PIC during the program. Unless they are tied to a definite value all disconnected inputs will float up and down and can cause random switching.

We need a clock signal and the simplest and cheapest is just an RC combination. The resistor should be between 5 kΩ and 100 kΩ and the capacitor should be greater than 20 pF. The values chosen for this circuit were not selected carefully and are not critical. In practice, it is difficult to predict the operating frequency of an RC combination. If we need a particular frequency it is better to use a variable resistor and adjust it to give the desired frequency.

Later on, when we have written the program we must tell the assembler what clock signal we are going to use. This programs the PIC to expect an RC oscillator. This information can, as an alternative, be included into the program. It's our choice but it is slightly easier to do it at the assembly stage but we must remember to do it.

The software

There are three layers of information that we need to use the registers. Firstly, we need the names and addresses of the files as we met in the register file map in Figure 15.8. The second step is the function of each bit in each register, and this is shown in Appendix A, and finally, an explanation of these functions as we see in Appendix B.

All this data looks daunting but the thing to remember is that we only use one bit at a time and we can ignore the rest.

All PICs and indeed all microprocessors and microcontrollers use binary code and nothing else. To make life easier for us, we use assembly language or a high level language. To use these languages, we need a program which is able to convert them to binary code. For assembly language, the required program is called an assembler and for higher level languages a similar job is done by a compiler.

An assembler is part of the PICSTART PLUS development system but there are other assemblers available that will handle code for all the PICs. The code that we write as the input to the assembler is called 'source code' and the code that is supplied by the assembler is called 'object code'. The object code is really in binary but to make it easier for us, it is usually displayed on the screen in hex.

Important note

As we type in the program we must not leave any spaces in instructions or in the data.

The programming steps

The purpose of our first program

We want to do something really simple but not something that may have happened by accident. We are going to configure all of PortB as outputs and then make the output signal to be off, on, off, on, off, on, off, on. We can connect a series of LEDs to the output pins so alternate lights will be on.

1 We will start by selecting Bank 1 of the register file map. To do this we need to talk to the status register, so by referring to Figure 15.8, we can see that the status register is file 03. From our look at the Status register in Chapter 15, we can see that bit 5 is the RP0 which controls the selection of Bank0 or Bank1. The first line of the program must set bit 5 of register 3.

Program starts:

        BSF   3,5   ; Sets bit 5 of register 3 to select Bank1

It is worth mentioning at this stage, that anything written after the semicolon is just a note for us and is ignored by the assembly program. This is called the comment field. It is always worth using this area to explain the program, and this often saves hours trying to remember what we were attempting to do when we first wrote the program. This can be an even larger problem if someone else writes a program and is then off sick and we are left to find out why the program doesn't work.

2 We now want to arrange for PortB to be an output. This involves loading a 0 into each of the controlling bits in the data direction register which we can see from Figure 15.8 is called TRISB and is register number 86. To do this, we are going to clear the W register using the CLRW instruction and then copy this zero into register 86. This involves two extra instructions so, at this stage, our program will read:

        BSF   3,5   ; Sets bit 5 of register 3 to select Bank1
        CLRW   ; puts a zero into register W
        MOVWF 86   ; copies the zero into register 86 which is
                         ; the PortB data direction register

3 Now we have sorted out the direction of the data flow and we can input the actual data. This time we are going to clear bit 5 of register 3 to allow us to access Bank0. We can now load our data into the W register. But what is the data? If we assume the LED is going to be connected between the port output and the zero volt connection, the voltages corresponding to the light sequence off, on, off, on, off, on, off, on will be 0V, +5 V, 0 V, +5 V, 0 V, +5 V, 0 V, +5 V and the data will be 0,1,0,1, 0,1,0,1. We could enter this as a binary number written as

B'01010101' or as the hex number 55 which is a little easier to read.

Once the 55H is in the W register, we can use the code MOVWF to copy it into register 06 which in Figure 15.8, we can see is the PortB data register.

The program is now:

```
BSF      3,5    ; Sets bit 5 of register 3 to select Bank1
CLRW            ; puts a zero into register W
MOVWF 86        ; copies the zero into register 86 which is
                ; the PortB data direction register
BCF      3,5    ; clears bit 5 of register 3
MOVLW 55        ; this is the output data to give the on, off
                ; sequence
MOVWF 06        ; this copies the data into PortB
```

4 As it stands, the micro will perform each of these steps once and then stop. We have a problem here because it will take only a few microseconds to complete these instructions – certainly too fast for us to see if the correct sequence of LEDs are illuminated. We need to give the micro something to do which will keep the LEDs operational and our choice here is to reload the output port continuously. Have a look at our new program:

```
      BSF      3,5      ; Sets bit 5 of register 3 to select Bank1
      CLRW              ; puts a zero into register W
      MOVWF 86          ; copies the zero into register 86 which
                        ; is the PortB data direction register
      BCF      3,5      ; clears bit 5 of register 3
      MOVLW 55          ; this is the output data to give the on,
                        ; off sequence
again MOVWF 06          ; this copies the data into PortB
      GOTO  again       ; this line forces the micro to return to
                        ; the previous line
```

The words 'again' are called labels and the assembler program notices that the two are identical and replaces them by the correct address. The fact that we have used 'again', a word that makes sense in the context is just to help us to understand the program, the assembler would accept 'asdf' or anything else just as happily. Some assemblers put restrictions on the names chosen. It may, for example, not allow it to start with a number, or use certain words or symbols.

5 At the end of the program, we have to put the instruction END to tell the assembler to stop. It is called an 'assembler directive' and is there to tell the assembler program that it has reached the end of our program. Directives are not instructions to the microcontroller and are not converted to machine code.

6 At the start of the program we can use the directive ORG which means 'origin' and gives the starting address for the assembled program. This has been added to our final program. If we had not done this, the assembler will assume the address to be zero so, in this case it would make no difference whether we added this directive or left it out. When the PIC is reset, it always goes to address 000 so if we wanted the program to start elsewhere we would have to leave a GOTO instruction at address 000 to tell the microcontroller where it is to start. Remember that the ORG is only an instruction to the assembler telling it where to start loading the program – the PIC doesn't know anything about this because directives are not converted to the program code.

So our final program is:

```
          ORG     000     ; Program starts at address 000
          BSF     3,5     ; Sets bit 5 of register 3 to select Bank1
          CLRW            ; puts a zero into register W
          MOVWF   86      ; copies the zero into register 86 which
                          ; is the PortB data direction register
          BCF     3,5     ; clears bit 5 of register 3 to select
                          ; Bank0
          MOVLW   55      ; this is the output data to give the on,
                          ; off sequence
again     MOVWF   06      ; this copies the data into PortB
          GOTO    again   ; this line forces the micro to return to
                          ; the previous line
          END             ; the end of our code to be assembled
```

Notice how the program is written in columns or 'fields'. It is necessary to use the correct fields as this tells the assembler what the items are. Remember to use the semicolon to start notes that we wish the assembler to ignore.

Connecting the LEDs

For clarity, only one LED is shown but an LED and resistor should be joined to all the pins 7–13 to show the full output.

LEDs come in different colours and sizes and the cathode must be connected to a less positive voltage than the anode. The cathode is generally recognized either by a shorter connector wire or a flat moulded onto the body.

Component values

Looking at the data for a standard red LED, the typical voltage (V_f) across them when lit is 2 volts with a maximum current of (I_f) 20 mA. The small 'f' stands for 'forward'. The light lost by reducing the current below its maximum value is not very great and it would be quite reasonable to operate the LED on, say, 10 mA.

To limit the current flow, a resistor is connected in series. Now, if the supply voltage for the microcontroller is 5 volts and about 0.7 volts are 'lost' inside the PIC and the LED is using 2 volts, the series resistor must be 'using up' the other 2.3 volts. The value of the resistor is given by $R = V/I = 2.3/(10 \times 10^{-3}) = 230\,\Omega$. If in doubt start with 470 ohms and see how it goes – this is a generally safe value for all situations.

More labels

The use of labels not only makes the program more readable but it allows modifications to be accommodated. For example, if we put in the actual address instead of the label and then modified the program by adding an extra instruction, the actual addresses would all shuffle along a bit to make room for the new instruction, making our old address inappropriate. The program would not work and it might take us hours before we see what we have done whereas a label would be sorted out by running it through the assembler with the new instruction added.

There is another useful assembler directive, EQU, which is an abbreviation for equates or 'is equal to'. This can be used to make programs more readable by replacing some of the numbers with words. For example, register 86 is the PortB Data Direction register but the program would be easier to read if we replaced the number by the name. This would be done adding the line: PortBDDR EQU 86 before the program listing so as soon as the assembler spots the name PortBDDR it would replace it with 86. This has no affect on the final program but it makes life easier for us – which has got to be a 'good thing'.

If we add some other labels, the final program can now be written as:

```
;EQUATES

PortBDDR   EQU   86          ; PortB data direction reg.
                             ; is register 86
PortB      EQU   06          ; PortB data register is
                             ; register 06
Status     EQU   03          ; Status register is register
                             ; 03
RP0        EQU   05          ; Bank1 is selected by bit 5
Data       EQU   55          ; Data used is 55H
           BSF   Status,RP0  ; Sets bit 5 of register 3 to
                             ; select Bank1
```

```
            CLRW                    ; puts a zero into register
                                    ; W
            MOVWF   PortBDDR        ; copies the zero into
                                    ; register 86 which is the
                                    ; PortB data direction
                                    ; register
            BCF     Status,RP0      ; clears bit 5 of register 3
            MOVLW   Data            ; output data to give the
                                    ; on, off sequence
again       MOVWF   PortB           ; this copies the data into
                                    ; PortB
            GOTO    again           ; this line forces the micro
                                    ; to return to the previous
                                    ; line
            END                     ; the end of our code to be
                                    ; assembled
```

By making full use of labels, we have rewritten our program without any numbers at all. This is just a matter of choice – all labels, some labels or no labels, whatever we like.

Using a crystal

This gives a more accurate clock speed so that programs that involve real can be written. It may be that we want a display sequence to run at a particular rate.

To change to a crystal we need to set up the configuration bits in the PIC so that it knows that it is being controlled by a crystal instead of the RC method. This is most easily handled during the assembly process by clicking on 'configuration bits' and selecting the clock source from the options offered.

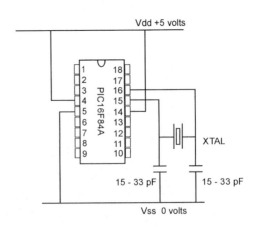

Figure 16.3

A Crystal
Controlled
Clock

The two capacitors shown in Figure 16.3 in the clock circuit always have equal values and the range shown is suitable for 200 kHz and all clocks of 2 MHz and over. Other recommended values are: 32 kHz – 68/100 pF and 100 kHz – 100/150 pF. The higher values in each range results in higher stability but slower startup times.

A ceramic resonator can be used as a plug-in replacement for the crystal.

A modification to the program

In the last program we controlled the voltages to each of the PortB outputs. With slight modifications we would be able to apply any combinations of voltages to control any external circuits. Even this first circuit has significant control capabilities but now we are going to extend the capability by applying a counting sequence to the output signals.

All programs are built on the backs of other programs that we have used before so we can save considerable time by keeping copies of our successful programs to be recycled whenever possible. This is well demonstrated in this example.

The program consists of three steps, two of which we have already designed and tested, so we know it works. If the new program refuses to work, we don't have to start from scratch, we know two–thirds of it is OK. This is a very powerful method of designing programs and whole libraries of programs are available so new developments can be reduced to slotting together ready-made program segments.

When we make changes to a previously program, it is important to save the new version under a new name so that, in the event of a disaster, we can retreat and start again.

Here is the section that we have 'borrowed' from our previous work:

```
ORG     000
BSF     3,5
CLRW
MOVWF   86      ; PortB data direction = output
BCF     3,5
MOVLW   55
MOVWF   06      ; PortB data set to a start value
```

At this stage we can, of course, set the start value for the output to any value between 00H to FFH which is binary B'00000000' to B'11111111'.

We have only one new instruction to worry about: INCF f,d. It increments or increases the value of a selected file 'f' by 1, and where the new value goes to is determined by the value of the 'd' term. If 'd' is 0 the new value is put into the W register but if it is 1, the new value is put back into the register in use.

PortB data register is register 06 so the code INCF 06,1 will take the current value of PortB data, increase it by 1 and put the answer back into PortB data so our starting value of 55 will change to 56 and the output voltages on the pins will change from 01010101 to 01010110.

This was just a single count, but for a continuous count we could use a label to make the program jump back and do the INCF trick again and again. When it reaches its maximum value, it will roll over to zero and start again so the count process can be continuous.
Our program would now be:

```
        ORG         000
        BSF         3,5
        CLRW
        MOVWF       86      ; PortB data direction = output
        BCF         3,5
        MOVLW       55
        MOVWF       06      ; PortB data set to a start value
again   INCF        06,1
        goto        again   ; go back and INCF again
        end                 ; end of source code
```

One more step

The speed at which the count continues is determined by the rate at which instructions are being followed.

Slowing things down

If we wish to slow things down, we can give the microcontroller something to do just to keep it busy. We have a NOP instruction which does absolutely nothing but takes one instruction cycle to do it. Since it doesn't do anything, it doesn't matter how many we include in a program, or where we use them. For a significant delay we made need hundreds, which is not an elegant way of solving a problem.

In the last modification to the problem, we made it count up on the PortB register. Now this takes time, so we could use this counting trick as a time waster. The PIC has 68 general purpose registers that can be made to count for us. Just choose any one of them and have it count for a set number of counts and then we can go back and count once on the PortB register, then go back to the time waste count. In Figure 16.4, we have loaded a register with a number, say

30H (48 in decimal). The next instruction decreases it by 1 to give 2FH (not 29!!!) and since the answer is not zero, we go around the loop and decrement it again to 2EH and so on until it gets to zero whereupon it leaves the loop to carry out 'instruction 2' shown in the figure.

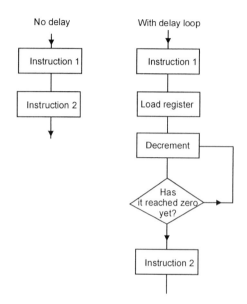

Figure 16.4

Using a timing loop

The instruction we are going to use this time is INCFSZ f,d. This is designed just for this type of counting job. It decrements the chosen register and if d = 0, the result goes into the W register but if it is 1, it will go back into the same register. For our purposes we would load the code as DECFSZ 20,1. This would decrement register 20 and put the answer back into register 20. When this register reaches zero, it will miss out the next instruction to stop it going around the loop again and will move on to the next instruction.

A slower count

Once again, this uses some of our previous programs.

```
          ORG      000
          BSF      3,5
          CLRW
          MOVWF    86      ; PortB data direction = output
          BCF      3,5
          MOVLW    55      ; PortB data set to a start value
          MOVWF    06
again     INCF     06,1
```

229

```
        MOVLW      30      ; Loads W with 30H
        MOVWF      20      ; puts the number 30 into file 20
count   DECFSZ     20,1    ; decrements register 20
        goto       count   ; keeps decrements until it gets to zero
        goto       again   ; returns to increment PortB
        end
```

For the slowest count on PortB, we would have to increase the count number in register 20 to its maximum number which, using this microcontroller is 7FH or 127 in decimal.

Calculating the delay

Starting from the moment that PortB is incremented:

MOVLW takes 1 count.

MOVWF takes 1 count.

DECFSZ takes 1 count normally but 2 when it leaves the loop.

As the register was loaded with the hex number 30, which is 48 in decimal, it will go around the 'count' loop 47 times at 1 instruction clock each and 2 clocks as it leaves the loop. This gives a total of 49 cycles.

goto will be used 48 times at 2 clocks each giving a total of 96 clocks.

goto will also be used once to return to the PortB, this is another 2 cycles.

Finally, INCF takes 1 count to increment the value on PortB.

The total is: 1 + 1 + 49 + 48 + 2 + 1 = 102 cycles

Assuming a crystal frequency of 32 kHz, we can divide it by 4 to give the instruction clock frequency and then by the delay of 102 cycles to give the rate at which the PortB is incremented of about 78 counts per second. PortB counts in binary from 0000 0000 to 1111 1111 and will finish its count after 256 counts so it will start recounting after 256/78 or roughly 3.3 seconds. We could reasonably double this time delay by a liberal sprinkling of NOPs or using a longer loop.

Longer delays

We have three alternatives.

1 For small changes, we could add some NOPs inside of the counting loop to boost the number of counts.

2 Our delay was built into the main program but we could have used it as a subroutine. A subroutine is any block of code that we may want to use more than once. In the main program we insert an instruction CALL followed by a label to identify the block of code so for our delay loop which we called 'count' we would insert the instruction 'CALL count' at any time we want to use our program to cause a delay. When the delay loop 'count' has been completed, we insert the instruction RETURN at the end of this block of code and the microcontroller will return to the main program.

The benefit of using a subroutine is that we can run the 'count' delay twice just by inserting the instruction CALL count twice in the main program and we don't have to enter the delay loop again with the fear that we will mistype something and it will all collapse. We can make a subroutine as long as we want and use it as often as we want just by adding the CALL and RETURN instructions.

Here is our previous program but reorganized to use the delay loop 'count' as a subroutine.

```
count   DECFSZ    20,1   ; decrements register 20
        goto      count  ; keeps decrements until it gets to zero
        RETURN

        ORG       000
        BSF       3,5
        CLRW
        MOVWF     86     ; PortB data direction = output
        BCF       3,5
        MOVLW     55
        MOVWF     06     ; PortB data set to a start value
again   INCF      06,1
        MOVLW     30     ; Loads W with 30H
        MOVWF     20     ; puts the number 30 into file 20
        CALL      count
        goto      again  ; returns to increment PortB
        end
```

The subroutine is called count and has the instruction RETURN at the end.

The main program has the instruction CALL count which means 'go and get a subroutine and use the one called count'.

We can then put:

CALL count
CALL count
CALL count

In the main program which would be an easy way to treble the length of a delay. We could design a subroutine called '1second' and another for '0.1second'.

Then if we needed to insert a delay of 2.3 seconds, we could just add:

CALL 1second
CALL 1second
CALL 0.1second
CALL 0.1second
CALL 0.1second

All subroutines would end with the same code RETURN, so how do they know where they have to go back to?

The answer is a series of memory locations called a stack. Each return address is stored in the stack in order that each CALL occurs, the relevant address is sent to the stack and as each RETURN will occur in sequence, the addresses will be unloaded from the stack in the order required. This is a first-in last-out (FILO) organization. See Chapter 8 for more on the stack.

A subroutine can include a CALL to another subroutine. These are called nested subroutines – the PIC16F84A has room in its stack for eight return addresses – which is pretty small by microprocessor standards.

3 In the PIC, most instructions are completed in a single instruction cycle which is $\frac{1}{4}$ of the clock speed. To change the delay, we could always change the clock speed. There are two benefits, a slower clock speed reduces the power consumed, there is no low-speed limit for the PIC, unlike some devices. Generally subroutines are preferred as there are often other constraints on the clock speed.

Quiz time 16

In each case, choose the best option.

1   Return is:

(a) only used as part of a delay loop.
(b) a ticket to take you home again.
(c) an assembly directive.
(d) an instruction found at the end of a subroutine.

2   ORG is:

(a) never needed since the PIC always starts at address 0000.
(b) an assembler directive.
(c) short for orgasm.
(d) an instruction code.

3   An assembler converts:

(a) decimals into hexadecimals.
(b) main codes into subroutines.

(c) source code to object code.

(d) object code into binary code.

4 In choosing a clock circuit:

(a) a ceramic resonator is not as accurate nor so robust as a crystal.

(b) an RC runs at four times the frequency of a crystal.

(c) a crystal gives the most accurate and stable frequency.

(d) use an RC circuit and a crystal to get accuracy and robustness.

5 The normal execution time for when using 4 MHz crystal is:

(a) 0.25 microseconds.

(b) 1 millisecond.

(c) 4 milliseconds.

(d) 1 microsecond.

17

Interfacing

Interfacing is the process of connecting a microprocessor to the rest of the circuit or to external devices. Even in the simplest of computer systems, there is some input device like a keyboard. So how does the microprocessor know that we have pressed a key? When we send text to a printer, how does the printer tell us that it is ready for more input?

In a general purpose microprocessor-based system, if it is to do anything useful, there must be inputs and outputs. The external devices must therefore communicate with the microprocessor. In some cases, the microprocessor takes the matter into its own hands and sends data out as part of its program but even in this case it normally allows the external device to help.

If a microprocessor-based system were used to heat some water, it is easy enough to imagine the program switching on the power supply and sitting there doing nothing for 10 minutes. It would be a better use of the microprocessor to leave the heater running and wait for a thermostat to signal that the water has reached the required temperature. This thermostat signal would arrive at an interrupt pin on the microprocessor.

Interrupts

Interrupts were introduced in Chapter 8 when we looked at the operation of the interrupt flag in the status or flag register but we will now delve a little further into the system.

All microprocessors have interrupts that can be initiated either by the software being run at the time or by external hardware circuits. Microprocessors differ in the details of their response to hardware interrupts and in the number of different interrupt pins offered. Details are always itemized in the technical data supplied with the device.

The likely options are as follows

There are two basic types of hardware interrupt. The first is an interrupt request or IRQ (or INTR) pin. This tells the microprocessor that it would like to have some attention. Since it is a request rather than an order, the microprocessor is free to say 'yes' or 'no' or 'yes, but not at the moment – just wait till I'm ready'. This does not imply any intelligence on the part of the microprocessor – it must be told which response to give by the software that is being run at the time. In the absence of any instruction, it normally accepts the interruption. If a particular interrupt pin has been told not to respond to an interrupt, we say that the interrupt has been 'masked'.

The second type is called a non-maskable interrupt, that is, an unstoppable demand for attention. This will always assume top-priority. A typical use for this would be for an emergency shutdown in the event of a power failure. The program can also instigate an interrupt by means of a software instruction as part of a program.

What is a hardware interrupt signal?

This is a change of voltage on an interrupt pin generated by the external device. The change required would be detailed in the technical data but there are four choices.

The first two choices are changes of voltage level. The pin can sit at +3.3 V or whatever the 'high' voltage happens to be, then responds when it falls to 0 V. We call this 'active low'. When the pin goes low, an interrupt is recognized. Alternatively, it could sit at 0 V and become activated by an increased voltage. This we call 'active high'.

Figure 17.1

Four ways of signalling an interrupt

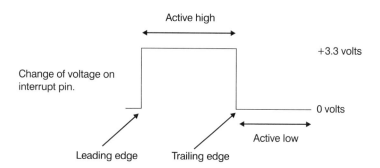

Active high

Change of voltage on interrupt pin.

+3.3 volts

0 volts

Active low

Leading edge   Trailing edge

The alternative approach is to use the sudden change of level. From low to high is called 'leading edge' or 'rising edge' and from high to low is the 'trailing edge' or 'falling edge' (See Figure 17.1 and have a glance at Figure 6.4 to see the alternative names.)

Once an interrupt is activated, the signal must be returned to its normal voltage before it can be triggered again. The input is masked as the interrupt program is running to prevent the interrupt pin from interrupting itself if the voltage remains at its active level.

Accommodating several external devices

The simplest and quickest way of connecting devices to the interrupts of a microprocessor is to have each device connected to its own interrupt pin. This is OK providing there are enough pins. Few microprocessors have more than two interrupt pins so we have to connect several devices to the same pin.

When an interrupt occurs, a program called the 'first level interrupt handler' or FLIH is activated. The function of the FLIH is to identify the device causing the interrupt and to pass the controls over to the 'interrupt handler' or 'interrupt service routine' program that has been written to deal with that device.

How does it know which device is crying for help? There are two options, polling and vectored interrupts.

Polling interrupts

This is a slow but sure way. Each possible device is interrogated in turn with an 'is it you?' signal until the source of the interrupt is found. The order of checking is prioritized so the most important device is checked first.

Vectored interrupts

Immediately after the interruption has occurred, the FLIH puts out a 'who's there?' signal and the interrupting circuit puts an identification signal onto the data bus. This signal is usually used as part of an address to identify the section of program to be executed.

What happens if an interrupt is received while the previous interrupt is being dealt with? Again we have a choice. We can disable the new interrupt until the first one is complete, then deal with the new one. Alternatively, we can check the priority of the new alarm and decide on the new priorities. A higher priority causes the present interrupt to be halted and its current state to be saved in the stack while the new one is worked on and then we return to unload the stack information and carry on with the original problem. If the new interrupt is less important it gets a 'wait a bit' message until the first one is finished.

In a microprocessor-based system it is up to the designer to decide on the priorities and uses of the interrupts. In a PC, the interrupts are prioritized in order of speed and importance. Top priority is given to the internal clock. This is the clock that tells us the time – not the square wave clock that synchronizes the circuitry. Thereafter, in order, we have the keyboard, two spare ones for any other operations, then comes the serial port, the hard drive, the floppy disk drive and finally the printer.

Parity

If you were to walk into a crowded room and say 'Burgers' and nothing further, many of those present would turn to their neighbours and say 'What was that?' and some would just stand and stare. (Some may even mis-hear and feel offended.) As a form of communication, this is not very efficient. Try instead, walking in and saying 'Lunch will consist of burgers'. Everyone would understand your message.

The first attempt was very efficient in terms of the number of words used but is probably likely to be inefficient communication since many people will not receive the message. In the second attempt, we have used five words to make sure that the one important one gets through. This is called adding 'redundancy'. The more redundancy we add, the more certain is the message but the slower and less efficient becomes the communication system. Data being returned from space probes use very high levels of redundancy, over 96%, which allows for correction of really scrambled signals due to the extremely low power levels involved.

We can use parity for alerting us to the possibility of an error in a stream of data or, in some cases, we can detect and correct the error. In its simplest form, we take a group of bits in a transmission, 4 or 8 bits are normally used though the idea is applicable to other values. In this example, we will look at a 4-bit group, say 1001. At the transmitting end, we add an extra bit on the end, either a 0 or a 1 to make the total number of '1's an even number. In this case, there are two '1's and so the number is already even, so we add a zero. The data now reads 10010. At the receiving end, if the data has been mutilated and now reads 11010, a quick count will show that there is an odd number of '1's and so an error has occurred.

This simple approach can be easily fooled. If there are two errors there will be an even number of '1's and passed as correct. And, another disappointment, if it shows an error, we cannot tell which bit is wrong and therefore cannot correct it. When this system is used, an error signal is sent back to the transmitter requesting a repetition but this

237

assumes that the transmitter and the receiver are in communication with each other. We can modify the system to make limited automatic correction feasible.

Let's assume we have, say, 16 bits of data to send.

Step 1 **Rewrite the data in the form of a square:**

```
0  0  1  0
1  1  1  1
0  1  0  1
1  0  1  1
```

Step 2 **Add parity bits. Across the top row, we have the numbers 0010 which includes a single '1'. In this system, which we will call 'even' parity, we add another '1' if necessary to ensure that there is an even number of '1's across the first row. As we have only a single '1', we add another '1' on the end. It now looks like this:**

```
0  0  1  0  1
1  1  1  1
0  1  0  1
1  0  1  1
```

The top row now has an even number of '1's. The next row has four '1's which is an even number so we do not need to add another '1'. We therefore add a '0':

```
0  0  1  0  1
1  1  1  1  0
0  1  0  1
1  0  1  1
```

The third row will be completed with a '0' since it contains an even number of '1's and the last row, with three '1's will need an extra one to be added. The result is now:

```
0  0  1  0  1
1  1  1  1  0
0  1  0  1  0
1  0  1  1  1
```

Step 3 **We now have five columns down the page and we can add extra '1's in the same way to make the total number of '1's an even number. The first two and the last columns each contain two '1's so zeros will be added. The third and fourth columns have three '1's so we need to add an extra '1' to each. The result is now:**

```
0  0  1  0  1
1  1  1  1  0
0  1  0  1  0
1  0  1  1  1
0  0  1  1  0
```

Notice how we have now got a total of 25 bits to be transmitted. This represents 16 bits of data and 9 bits added to check the accuracy of the data. The final serial transmission is 0010111100101010111100110. This means that 9 out of 25 or 36% of the transmission is not actual data and represents redundancy.

Let's see how it works. We will assume an error has occurred and one of the bits is received incorrectly so here is the received transmission:

0010111110010101001100110

Step 1   Layout the data as a 5 × 5 square.

```
0  0  1  0  1
1  1  1  1  0
0  1  0  1  0
1  0  0  1  1
0  0  1  1  0
```

Step 2   Check the parity in each row across the square.

We decided to make each row and column to have an even parity.

The first row has two '1's, this is even – OK.
The second row has four '1's, this is even – OK.
The third row has two '1's, this is even – OK.
The fourth row has three '1's, this is odd – an error has occurred.
The last row has two '1's, this is even – OK.

We now know that one of the bits in the fourth row has been received incorrectly.

Step 3   Do the same for the columns.

The first column has two '1's, this is even – OK.
The second column has two '1's, this is even – OK.
The third column has three '1's, this is odd – an error exists in column three.
The fourth column has four '1's, this is even – OK.
The last column has two '1's, this is even – OK.

Step 4   Isolate the error and change the data.

The error occurs in the third column and the fourth row. Since this is now known to be an error and we only have a choice of 0 or 1, we can confidently change the 0 to a 1 and recover the correct data stream.

```
0  0  1  0  1
1  1  1  1  0
0  1  0  1  0
1  0 [1] 1  1
0  0  1  1  0
```

In this example, we chose to use even parity, that is, we made each row and column have an even number of '1's. It would work equally well if we used odd parity by making the number of '1's an odd number. It would also work just as well if we counted the zeros instead of the ones. If more than one error occurs, it will warn us of an error but it will be unable to make any corrections. If you try it, you will see that it indicates four possible positions for the two errors and nine for three errors.

Example

Correct this received data which includes one error. To provide automatic correction, odd parity on the '1's has been used.

The received signal: 1101001001111001010101101.

Step 1 Layout the data as a 5 × 5 square

```
1  1  0  1  0
0  1  0  0  1
1  1  1  0  0
1  0  1  0  1
0  1  1  0  1
```

Step 2 Check the columns and rows for an odd number of '1's

```
1  1  0  1  0 ✓
0  1  0  0  1 ✗
1  1  1  0  0 ✓
1  0  1  0  1 ✓
0  1  1  0  1 ✓
✓  ✗  ✓  ✓  ✓
```

Step 3 Isolate the error

```
1   1  0  1  0 ✓
0  [1] 0  0  1 ✗
1   1  1  0  0 ✓
1   0  1  0  1 ✓
0   1  1  0  1 ✓
✓   ✗  ✓  ✓  ✓
```

Step 4 The data where the column and row intersect is the error. So we simply change the '1' to a '0'.

Step 5 Strip out the parity bits to recover the original data

11010000111010100110

Data transmission

The most basic way of sending information from one place to another is simply to connect a wire to both ends of the system and apply a voltage to one end. By making the voltage vary, we can send different levels and even speech or music. These are called analogue signals and have many drawbacks.

The main one is the effect of noise. As the signal travels along a wire it gets weaker and it has noise induced into it by random electro-magnetic signals and vibration of the molecules of the conductor. The overall effect is that the signal becomes degraded and weaker.

The 'weaker' bit is no problem, we can soon amplify it back to its original size but the noise is a different matter. The electrical noise has become embedded into the signal and has permanently distorted it. Amplifying it will amplify the noise and the signal equally.

We have no problem with digital signals since we know that they will all be rectangular in shape and so the amplifier can be used to regenerate the shape of the signal and hence strip out the effects of the noise. If we are faced with sending something inherently analogue, like speech or music, our first job is to convert it to a digital form.

Analog to digital conversion (A to D or ADC or A–D)

Inevitably these days, this is taken care of by an integrated circuit of which there are many different designs. It is quite possible, but totally uneconomic, to construct our own ADC so it is really a matter of flicking through the catalogues and choose the most appropriate one available.

There are several different designs of ADCs, which are based on three basically different approaches.

Flash converter

The first is called a flash converter or parallel encoder. These use circuits called comparators. A comparator has two inputs, one is the analogue voltage being converted and the other is a known reference voltage.

All we ask of a comparator is to answer a simple question: 'Is the analogue input voltage higher or lower than our reference voltage?' It answers by changing its output voltage to a logic 1 to mean it is higher and a logic 0 to mean it is lower. They are so accurate that the chance of it accepting the two voltages as the same level are extremely slight and doesn't happen in practice (see Figure 17.2).

Figure 17.2

A comparator used in a flash ADC

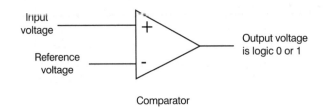

Comparator

If the input voltage is higher than the reference voltage the output voltage gpes to a logic 1 value.

So how do we use the comparator? If we had three of them with reference voltages set to 1 V, 2 V and 3 V and then applied the input voltage of 2.5 V to all of them, the first two would set their outputs to a level 1 and the last one would be unaffected at logic 0. The logic levels could be used to generate a binary number to represent 2.5 V.

An input voltage of 2.4 V or 2.9 V would also result in the same comparators being activated and hence the same output digital signal. This error occurs in all analog to digital converters. We can reduce the size of the error by increasing the number of comparators to detail more levels. Real ones have between 16 and 1024 different levels.

Ramp generators

These are a combination of a binary counter that simply counts up from zero to its maximum value, perhaps 1024 like the last type. As the binary count proceeds, a ramp voltage is made to steadily increase. A single comparator is used to compare the output of the ramp voltage with the analog voltage being converted. As soon as the ramp voltage exceeds the input voltage, the comparator signal stops the counter. The counter output is then the digital equivalent of the analog signal (see Figure 17.3).

Successive approximation

If we were to use a 3-bit digital signal to convert an analog voltage of between 0 V and 4 V we could have the 3 bits representing voltages of 4 V, 2 V and 1 V. This is how the circuit responds to an input of 3.5 V.

Figure 17.3

An ADC that uses a ramp voltage

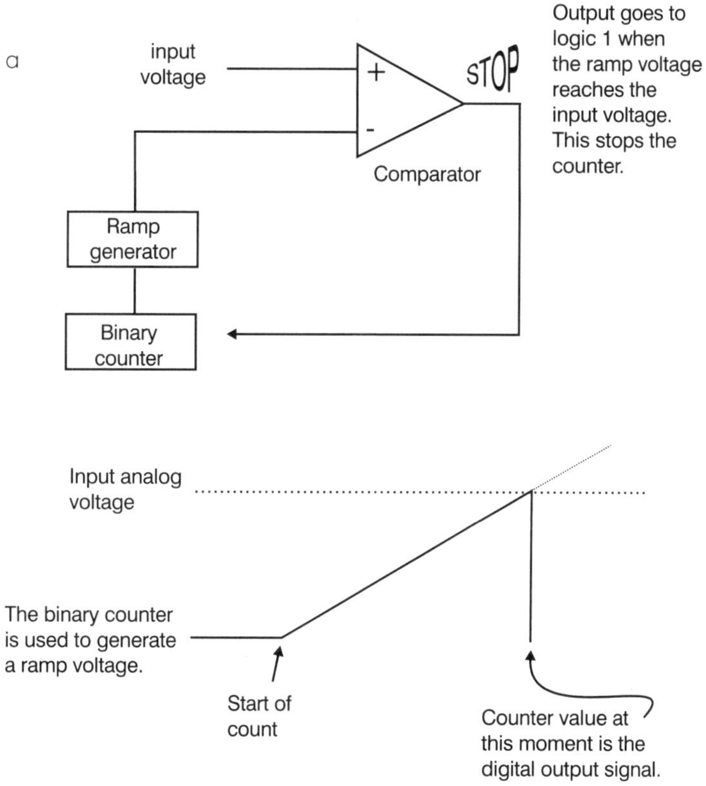

input voltage

+

STOP

−

Comparator

Output goes to logic 1 when the ramp voltage reaches the input voltage. This stops the counter.

Ramp generator

Binary counter

Input analog voltage

The binary counter is used to generate a ramp voltage.

Start of count

Counter value at this moment is the digital output signal.

The digits are initially set to 000. The left-hand bit is switched on and its 4 V is compared with the input. The input is seen to be less than this so this digit is reset to zero. It then tries the next bit and its 2 V are compared and found to be less than the input so it remains set. The digital signal is now 010. The circuit now adds the 1 V from the last digit. The result is a total of 3 V, which is compared with the input analog signal. The input of 3.5 V still exceeds the current value of 3 V so the last bit is set. The final digital output is 011.

The circuit has tried all the available values until it finds the one that provides the result closest, but less than the input signal. As before, the more bits we are using, the more accurate is the result.

In checking the specifications of likely ADCs to use, we need to compare the following criteria.

Quantization error

In the above example using the flash converter, we can see that an analog input of 3.5 V would provide the same output as would any value between slightly over 3 V and slightly less than 4 V. This error

243

means that small variation in the analog input voltage will be lost. The size of this error is equal to the space between the comparator reference voltages.

Changing from eight comparators to 1024, would mean that the voltage gaps would decrease from 1 V to 7.8 mV as would the quantization error. Regardless of the method used for A–D conversion, quantization error is always present.

Bits

The more bits, the merrier. Likely values will be between 8 and 16.

Speed

There are two factors here. How often can we get an updated value for the signal and how well can we follow any changes it is making? Even if we do not want the signal to be sampled at a very high rate, we still may want to take a quick sample so that the input value is unlikely to change very much as the sample is actually being measured.

For speed, you cannot beat the flash converter. It can sample for a period as short as 3 ns which compares very favourably with the typical values of 10 μs for the ramp generators and successive approximation types.

Digital to analog conversion (DAC)

Changing a group of digital bit values to an analog voltage is basically just the reverse process of the A–D conversion that we met in the previous section.

Most digital to analog converters operate by adding current together then converting the result into an analog voltage. The binary levels are used to switch currents on or off.

Let's assume a 4-bit digital signal in which the most significant bit is made to generate a current of 8 mA and the others produce, in turn, 4, 2 and finally 1 mA. If the digital signal to be converted happened to be 1011_2, then the first, third and fourth current sources would be activated giving a total of $8 + 2 + 1 = 11$ mA (Figure 17.4).

In some DACs the final output is a changing current but in others it has been converted to a variable voltage. It just depends on which integrated circuit you choose to use. In the ones offering a voltage output, the total current is then passed through a resistor. If we choose a nice easy value like 1 kΩ, the voltage across the resistor would be 11 mA $\times$ 1 kΩ = 11 V.

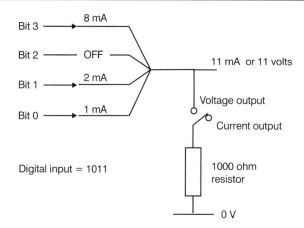

Figure 17.4
A DAC with a current or voltage output

Bit 3 — 8 mA
Bit 2 — OFF
Bit 1 — 2 mA
Bit 0 — 1 mA

11 mA or 11 volts

Voltage output
Current output

1000 ohm resistor

Digital input = 1011

0 V

In a similar way, we can see that all binary values between 0000 and 1111 would be converted into voltages between 0 V and 15 V. There are a couple of specifications that may need to be checked to decide on which one to use.

Resolution

This is the number of digital bits used to convert into an analog voltage. Typical values available are from 4 to 18 bits. As the digital input changes by a single bit, say from 1000 to 1001, the resultant voltage or current increases by a discrete step. The size of this step is determined by the number of bits used compared with the maximum value of the output current or voltage.

For example, if we used 4 bits then this would provide a total of 16 different steps and if the maximum happened to be 8 V, each step will represent a voltage change of 0.5 V. Thus, a steadily increasing digital signal will cause the analog voltage to increase in small discrete steps like a staircase. This is all very similar to the cause of quantization error.

Speed

The speed of operation is very dependent on the chip being considered. The conversion times available from an exceedingly fast 1 ns to a sluggish 5 μs.

Serial and parallel transmission

In sending information in digital form, we have a choice of using serial or parallel transmission. In the case of serial transmission, the binary values are represented by two different voltage levels and are sent one

245

after the other along a cable. This is simple but slow. The alternative is to have several wires and thus be able to send several bits of data at the same time, one on each conductor.

In a microprocessor-based system, even if it is a one-off, it is usually better to conform to established standards for the cabling so that other instruments and circuits can be connected with a minimum of hassle.

Parallel connection

There are several different standards used for parallel connection of data but one of the most widely used, and most reliable, was produced by Centronics.

The Centronics system sends eight bits at a time and employs a 36 plug and socket system. To send data, there are four basic control signals as well as the eight data lines. It also stipulates a variety of other control wires that can be used if required.

The important thing to remember about these standards is that you do not have to use all the connections listed but those that you do decide to use should conform to the stated specification and be on the correct pins. This ensures that if the plug is inserted into a new piece of equipment, it may not work but at least it will not be damaged.

Centronics data transmission

To see how the system works, we will use timing diagrams to show what happens and when. These diagrams, which all look very similar at first glance, are shown in all data manuals to show the sequence of events inside the microprocessor and in the surrounding circuit.

There are a couple of points that are worth mentioning. We have mentioned the problem of rise time in Chapter 7. You will remember that we cannot change a voltage level instantaneously. It may not seem an important delay when we think of switching a light on at home but when the microprocessor is handling data at millions of bits per

Figure 17.5

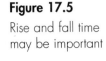

Rise and fall time may be important

An oversimplified version of a voltage pulse.

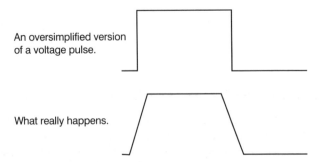

What really happens.

Figure 17.6

Showing alternative data levels

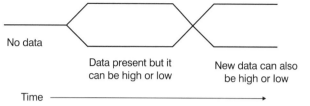

No data

Data present but it can be high or low

New data can also be high or low

Time

second, these delays can be important and is a common cause of failure in a circuit that 'should' work. Most waveform diagrams show the ends of a square wave as sloping lines rather than vertical ones.

In Figure 17.5 we have a positive-going pulse to represent a data value of 1 but, of course, data could equally well be at 0 V to represent a binary 0. In cases where we want to show that a level has changed, but it may go to either level, we redraw the diagram to show both possibilities at the same time, as in Figure 17.6.

Figure 17.7 shows the process of transferring eight bits of data from a microprocessor to an external printer or other device.

Step 1 The microprocessor puts the eight bits of data on the data wires.

Step 2 A short delay occurs while we wait for the data voltages to settle on all the eight wires. Then the strobe pulse occurs to tell the printer or other accessory that the data is ready. The line over the word strobe indicates that it is active low. No line would mean active high.

Step 3 The printer starts loading data and the busy line goes high to prevent more data being sent.

Step 4 When the data has been printed, the busy line goes down to tell the microprocessor to put the next piece of data onto the data wires.

Figure 17.7

The timing of Centronics signals

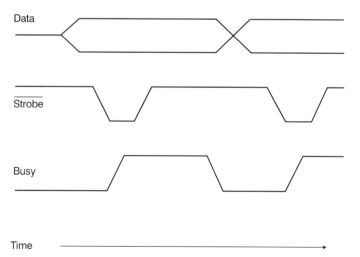

Data

$\overline{\text{Strobe}}$

Busy

Time

The standard allows for many other control wires for other purposes. In this example, the eight data wires have been used as outputs from the microprocessor but it is quite possible to use them to carry input data although this is not allowed for in the design of PCs.

Serial transmission

To send information in serial form requires only a simple communication link but it is inevitably slower than parallel transmission since the data is only sent one bit at a time.

UARTs

To convert the parallel data on the data bus of the microprocessor to a serial transmission we could use a shift register as in Chapter 6. The modern alternative is to use a chip called a UART (universal asynchronous receiver/transmitter) or USART (universal synchronous/ asynchronous receiver/transmitter).

These are integrated circuits that convert data from parallel to serial transmission so instead of having eight wires, each carrying a single bit at the same time, the serial transmission passes the bits, one at a time along a single wire. It can also receive eight bits of serial data then pass it into the microprocessor in a single byte. Parallel transmission is obviously a lot faster since eight bits are moved at a time but it requires eight connections. To send a fax signal, for example, would require the use of eight telephone lines whereas the UART can convert it to a serial transmission and sent it over a single line at one-eighth of the speed.

UARTS do a lot more than a shift register. They include parity checking and buffers to enable it to handle about 16 bytes at a time without involving the microprocessor. Most transmissions involve the ASCII code to represent the characters to be transmitted. This is a seven-bit code to represent each alphanumeric character and a variety of control instructions. For example, the letter E is 45 in hex which, using only the lower seven bits, is 1000101. The ASCII code is used in both parallel and serial transmissions. Each letter and symbol has its own seven-digit code. A further bit is added on the end to provide a parity bit or it can be used to swap over to an alternative set of characters to allow mathematical symbols and Greek letters to be transmitted or, if unused, can be left at zero. Our letter E would then be represented as shown in Figure 17.8.

When using ASCII signals in a serial transmission, we need to be able to tell the receiving apparatus when a particular ASCII character has been sent. This is easily done in a synchronous system that ensures that the transmitter and the receiver are locked together running at the same speed. This is not the easiest way of operating the system owing

Figure 17.8

Coding in ASCII

Bits 0   1   2   3   4   5   6   Parity

The ASCII code is 45H or 1000101 + the parity bit.
Notice that the bits are coded with the lsb first so
we actually send 1010001 + parity bit, in this case
using even parity.

to the difficulties of ensuring the two devices remain synchronous. Therefore, we tend to operate asynchronously. This means that we have to send a signal along with each ASCII code to tell the receiver when the code has started and when it has stopped. Otherwise the transmitter would send a continuous stream of data and if a bit were lost, the receiver would get out of step and would misread all subsequent data.

To get round this problem, a 0 V 'start' bit is sent at the beginning of the character and a positive 'stop' bit is sent at the end. This brings a seven-bit ASCII code up to a total of 10 bits. The start and stop bits ensure that there is at least one change of level for each character that can be used to keep the receiver clock nearly synchronized to the transmitter for the time taken to receive that character.

For distances over a few metres, we need to use a slightly more sophisticated transmission system to prevent random noise from interfering too much. There are several systems in use, the most popular being those created by the EIA (Electrical Industries Association).

As with most transmission media, there is a trade off between the speed and the maximum distance the system can be used for. If you intend pushing the transmission distance to its maximum value, you will have to accept a reduced speed. As a rule of thumb, halve the speed if you double the distance.

RS232C

This is one of the transmission standards created by the EIA committee. This standard allows for transmissions up to 50 feet (15 m) and at speeds of up to 20 kbaud (it can actually exceed this speed and distance but it's not guaranteed). The baud is the measure of the speed of transmission. It is the number of clock periods per second, which approximates to the number of bits per second.

The RS232C transmission is balanced at about 0 V. Here's the time to be careful, the binary one level is a negative voltage (between –5 and –15 V) and a binary zero level is a positive value between +5 and

249

+15 V. This seems upside down compared with all our previous uses of binary. Our letter E would be transmitted as in Figure 17.9. The transmitter levels are specified as ±5 V but the receiver limits are ∓3 V. This allows for a noise spike to be up to 6 V before there is any possibility of misreading a piece of data.

Figure 17.9

RS232C

transmission

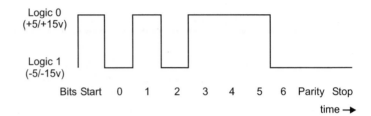

RS423A

This is an improved version having a maximum speed of 100 kbits/s and a maximum cable length of ¾ mile (1.2 km). The transmission voltages have to be between ±3.6 and 6 V and the receiver can go down to ±0.2 V.

Changing voltage levels

How do we change the binary or logic values into the RS232 voltage levels? If you are building a microprocessor-based system then the most obvious way is to use a pair of integrated circuits called the 1488 (transmitter) and the 1489 (receiver). These integrated circuits have been around for many years and are simple and reliable. They have a small snag in that they need 12 V supplies whereas nowadays 5 V supplies are much more common so you may find some new transceivers (made by Maxim) more interesting. These only require a single +5 V supply and generate their own ± voltages for the RS232C transmission. Each chip contains two transmitters and two receivers and operate up to 120 kbits/s. The devices are numbered MAX202, MAX208, MAX220 and MAX232 and others. PCs have a serial port that provides signals at RS232C levels.

Using RS232C in real life

Most RS232C links are via a 25-pin 'D' plug or a 9-pin 'D' plug and socket (Figure 17.10) but unlike the Centronics which is quite stable and usually work straight off, the RS232C can be a real nuisance. Before attempting to communicate, you must ensure that the transmitter and the receiver are using the same word length and parity values are set for the same speed of operation. Even then, it may take

Figure 17.10

'D' connectors for RS232C

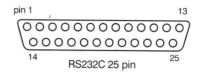

RS232C 25 pin

RS232C 9 pin

some experimenting before they spring into life. The problem is that there are many more options for the other connections. All have to be agreed between the receiver and the transmitter. The specifications are not detailed enough and can lead to different interpretations. It is not surprising that it is often insufficient to connect an RS232C cable between two pieces of equipment and switch on. You will need to get hold of the RS232C connection specification and settle down in a comfortable chair.

Modems

A modem (MOdulator DEModulator) converts a digital signal into two audio tones so that the transmission can occur along a telephone line. Telephones are generally designed to accept frequencies between 300 Hz and 3.1 kHz. This relatively narrow bandwidth was chosen to allow speech to be transferred with undue loss of quality while allowing the largest number of calls to be passed along the same cable. Once the digital signals are on a telephone line then the range is unlimited.

Choice of systems

A few metres

We can use the raw binary data transmitted over a simple cable (see Figure 17.11).

Figure 17.11

A very short range link up

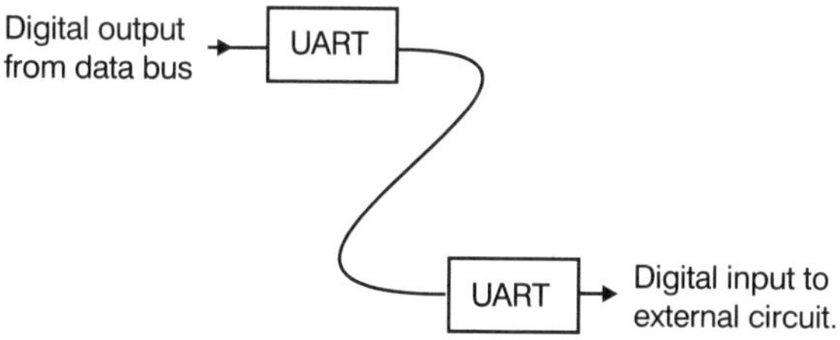

Digital output from data bus → UART

UART → Digital input to external circuit.

Tens or hundreds of metres

We can convert the transmitted signal to RS232C or RS423A as necessary (see Figure 17.12).

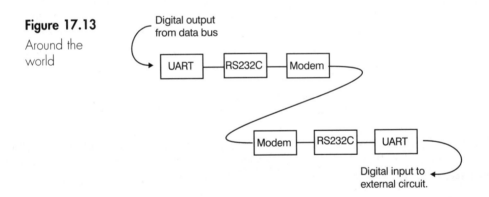

Figure 17.12

Around the building

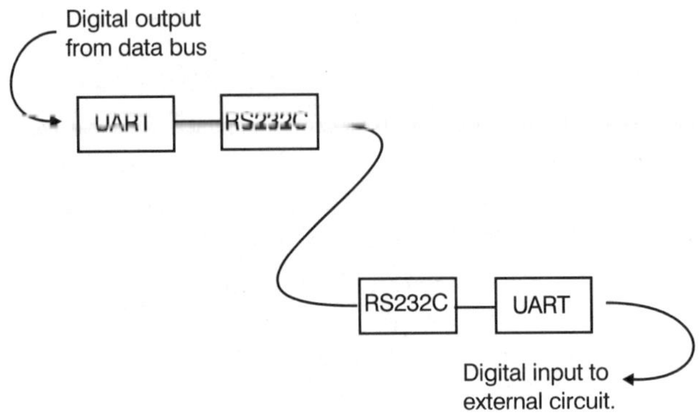

Unlimited range

Add a modem and link by telephone or optic fibre (see Figure 17.13).

Figure 17.13

Around the world

An optic fibre link

A piece of optic fibre is a solid piece of glass or plastic. The plastic fibre is about 1 mm in diameter and is suitable only for short ranges of a few tens of metres but it has the advantage of being cheap and easy to use. Its useful range is limited by the clarity of current plastics. The special silica glass is incredibly clear and hence has much lower losses and able to be used over any distance, with suitable repeaters. It also has a much smaller diameter – only about 125 μm before the external protective layers are added.

If a light is shone into the end of an optic fibre, it will reflect off the inner surfaces along the cable. The light source used is a laser operating in the infrared region of the spectrum. To use it as a means of sending a digital signal we need to switch the light source on and off and then detect the flashes of light at the far end of the cable by a photoelectric cell. The losses can be made up by repeaters just as we do on copper-based systems, so range of operation is no problem. The optic fibre does not suffer from any electric noise pickup along the route and has an enormous bandwidth. In one sense, it is not really optional because nearly all long distance telephone cables are now optic fibres. (See Further reading for our companion volume *An Introduction to Fiber Optics.*)

If we are constructing our own fibre optic link, all we need to do is to buy the laser (or light emitting diodes (LEDs)) and the photocells and some plugs and sockets to connect it all up. It can be used to replace the copper cable in any of the systems described (see Figure 17.14). Careful! The infrared light from the lasers can cause immediate and irreversible eye damage. We must always remember that we are down to our last pair of eyes.

Figure 17.14

A fibre optic link – any distance

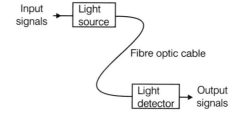

Data transfer rates

Using a single optic fibre for serial transmission, typical data transfer rates of 100 Mbytes/s are available up to 10 km. Very high speed data transfer of 1 Gbyte/s can be achieved up to 100 m using parallel transmission along a bunch of fibre optic cables.

Quiz time 17

In each case, choose the best option.

1 **The fastest design of analogue to digital conversion is a:**

(a) ramp converter.
(b) flash converter.
(c) comparator.
(d) successive approximation converter.

2 Quantization error can be reduced by:

(a) increasing the number of levels in the ADC.
(b) increasing the speed of the conversion.
(c) using vectored interrupts rather than polling.
(d) using a flash converter.

3 Using the RS232C standard, a binary 0 is most likely to be transmitted as:

(a) −4 V.
(b) +2 V.
(c) −5 V.
(d) +10 V.

4 A modem is:

(a) a type of USART.
(b) normally connected between the UART and the RS232C converter.
(c) only used in fibre optic systems.
(d) used to convert digital signals into audio tones for transmission over telephone cables.

5 This received transmission has sixteen bits of data and includes an error. It is using odd parity on the ones in a block of 25 bits: 0000111010111000100110011. The corrected data is:

(a) 1000111010111000101110011.
(b) 0000111010111000101110010.
(c) 0000111010111000101110011.
(d) 0000111010111000100110011.

18

Test equipment and fault-finding

This chapter is intended to give some pointers towards finding faults in a microprocessor-based system. This chapter is firmly based on experience and could equally well have been entitled, 'Mistakes I have made'.

What's gone wrong now?

The whole process of fault-finding should be undertaken slowly and carefully. There is a popular misconception that you have to keep busy, taking measurements, making adjustments and changing components. But, in fact, most of the time is spent just sitting and thinking (don't forget the last two words!).

Collect the symptoms and write them down. Be wary of other people's idea of the symptoms. If they have misunderstood what is happening you could waste hours or days going off at a tangent. If you forget to write them down, then sooner or later you will be back repeating the same checks.

Don't make the problem worse

In most cases, a piece of equipment or a circuit fails due to a single fault. Two simultaneous but unconnected faults are very rare. There are two popular ways of converting a small problem into a large one. These are static electricity and plugs etc.

Static electricity

When two different materials rub against each other, some negative electrons tend to migrate from one material to the other. This results in a voltage difference between the two materials. The amount of voltages can be very high – several thousand volts. If we walk across a carpet or sit on a plastic covered chair, we can become lethal to an integrated circuit designed for 5 V. Many integrated circuits have anti-static precautions built in but they have limited success. There is a trade-off here in that the better we make the antistatic precautions, the slower the integrated circuit can switch.

We can overcome the problem by reducing the build up of static by allowing it to leak away. In carpets, clothes and furniture we can do this by adding a wax or polish that absorbs and holds a small quantity of moisture. A slight dampness is a very effective way of preventing static problems. For this reason, the weather and air humidity is important. The death rate of integrated circuits tends to vary seasonally! It is not helped by air-conditioned plant where the humidity is low. The effect of static electricity on integrated circuits is difficult to predict. It generally causes small localized failures which can have very peculiar effects.

Better than spraying ourselves with water, we can take a more high-tech approach but how far to go in this direction depends on what is at stake. If we are going to handle a couple of cheap AND gates once a week, then only the simplest precaution is worthwhile. However, sitting on a production line plugging in microprocessors will make any precautions economic.

The simplest method is to have a conducting band clipped around your wrist with a lead going off to a ground (earthed) point. These wristbands are made of rubber into which carbon has been amalgamated to allow it to conduct slightly. As well as the wristband we can place a sheet of this rubber on the bench top and ground the bench. Such antistatic workstations are very effective. A word of warning. Do not make your own wrist strap from a length of copper wire. This offers a very low resistance and provides no protection against electrocution in the event of accidentally touching a power line.

At home, just avoid working on a plastic table or chair or wearing clothing made from man-made fibres. Natural materials like cotton, wool and untreated wood naturally absorb some water and are fairly safe. A nice wooden bench coated with polyurethane varnish is effectively a plastic bench and should be avoided.

Problems with plugs

Many plugs used between pieces of equipment have a large number of pins. Pulling one of these out with the power connected is going to disconnect some voltages before others. This can prove fatal for integrated circuits. Either all the supplies must be on, or all should be off so never plug or unplug anything with the power on. For the same reason, never remove or replace an integrated circuit with the power on.

Tests we can make without test equipment

Are the power supplies turned on? Do you need two supplies? If you are using two supplies, are they connected together to keep their voltages in step with one another? If a ground connection is required, is it connected?

Most power supplies have floating outputs. That means that a 5 V supply, for example, will have a 5 V difference between its two terminals but neither is connected to the ground potential. This means that if we connect the negative terminal to earth, as in Figure 18.1(a), the other terminal goes to +5 V. If, on the other hand, we make the connection shown in Figure 18.1(b), the other terminal will become −5 V.

Figure 18.1

Connecting floating supplies

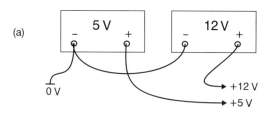

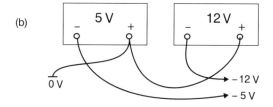

Have a look at the soldering if it is visible. It should be smooth and shiny. Any dull and craggy looking areas are suspect. If the integrated circuits are plugged into bases rather than being soldered, have a look to see if they have been inserted the right way round. Unfortunately, integrated circuit manufacturers take few precautions to prevent this type of error.

In most integrated circuits, the pins are numbered around the outside as shown in Figure 18.2. The position of pin 1 is always on the left-hand side of the end which has an indentation when viewed from the top as in Figure 18.2. When looking for the indentation don't be mislead by a small circular mark where the plastic has been molded. The printed circuit board usually has either a number '1' or a small square or other mark to indicate the position of the first pin.

Figure 18.2

Pin numbering of 'dual in line' (DIL) chips

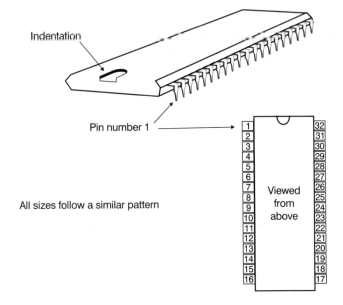

Indentation

Pin number 1

All sizes follow a similar pattern

Viewed from above

Figure 18.3 shows the pin grid array (PGA) layout. Notice that the letters skip from H to J because of the possible confusion between I and 1. The device determines the number of pins. The one shown happens to be the elderly Intel 80386. The Pentium has 21 pins along each side.

Simple test equipment

Apart from the standard voltmeter and an oscilloscope the only other simple piece of gear that may be helpful is the logic probe. It is better than the average oscilloscope at detecting very short voltage spikes and is faster to work with than a voltmeter.

Logic probe

A logic probe is a simple instrument that has two power connections and the other is a conducting tip that can be touched on points of interest. The general layout is shown in Figure 18.4. There are three LEDs on it. The first two show the logic states 0 or 1 and the third one indicates the presence of a high frequency square-wave or a single, very short duration, pulse, called a 'glitch'.

Figure 18.3

Pin numbering of
Pin Grid Array
(PGA)

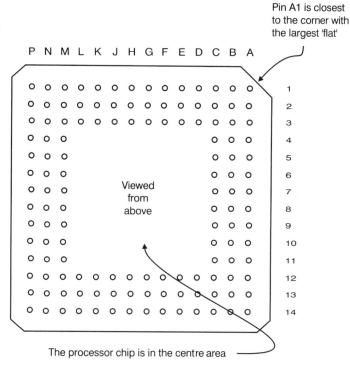

Pin A1 is closest
to the corner with
the largest 'flat'

Viewed
from
above

The processor chip is in the centre area

Figure 18.4

A logic probe

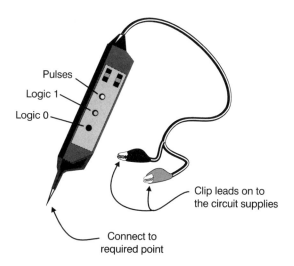

Pulses

Logic 1

Logic 0

Clip leads on to
the circuit supplies

Connect to
required point

259

Simple tests to make with these pieces of test gear

We can check some of the voltages on the microprocessor pins. If possible, it is a good idea to check on the actual pins rather than the base into which it is plugged. Doing this ensures that the base connections are also OK. It would also find the bent pin shown in Figure 18.5.

Figure 18.5

Pin bending not recommended

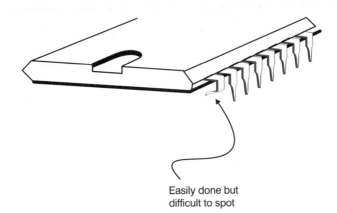

Easily done but difficult to spot

The likely pins that are worth checking are the ones carrying a dc voltage like the power supplies and the interrupts. It is worth keeping an eye on pins that should be at 0 V. When using a voltmeter, they can sometimes show 0 V when they are disconnected and floating. If you use your voltmeter to measure the voltage between the positive supply voltage and the suspect pin, it will still indicate 0 V showing that something is clearly amiss. A logic probe would not be fooled by a floating 'zero', it will not show a logic zero if it is floating.

The next job is to see if the microprocessor is running at all. We can do this by using the oscilloscope on a clock signal. Assuming that the clock signal is OK, we must next check that the microprocessor can follow an instruction and that the address and data bus are being read correctly.

A good check on the operation of the microprocessor can be arranged by getting it to do a simple repetitive program consisting of a permanent 'no operation' code. A no-operation code will instruct the microprocessor to do nothing except read the next instruction from the data bus by simply incrementing the value on the address bus. This new instruction will be another no-operation code and so the address bus will be continuously incremented. To provide a permanent no-op input we can solder or otherwise connect the required logic codes to the data bus. This is called hard-wiring the

Figure 18.6

The address bus counts up in binary

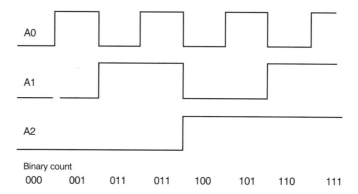

A0

A1

A2

Binary count
000   001   011   011   100   101   110   111

This only shows the first three address lines.

Notice how the frequency changes.

data bus. As the address bus counts up in binary the lowest address line will be switching rapidly between zero and one giving a square wave output.

If we look at Figure 18.6, we can see that line A1 is running at half the frequency of line A0. Similarly address line A2 has half the frequency of A1 and so on all the way along the address pins. If we connect an oscilloscope to each line in turn, the frequency should reduce steadily. Check for the halving of frequency on each address line and errors in wiring like short circuits between address lines will become apparent.

If we get this far and still things seem wrong, we are into serious fault-finding.

A serious piece of test equipment

All the previous pieces of test gear have failed when we try to see what is happening on the address and data buses under real operating conditions. The oscilloscope cannot watch more than two different places at the same time but we may need to monitor a larger number, perhaps 50 or more places and then slowly check the information back or print it out. An instrument called a logic analyser can achieve all these functions and much more.

It can answer such questions as:

○ What values actually appear on the address bus when we cause an interrupt to occur?
○ Is the correct program actually being run?
○ Are there any unwanted voltage spikes occurring?

The design of a logic analyser is basically a very simple combination of shift registers. You may remember we looked at shift registers in Chapter 6. The register was loaded with data and, on each clock pulse, the data is moved one place to the left or right as required.

Now imagine a shift-right register that can hold 36 bits of data. If we connect it to A0, the first line of the address bus, and run a program, the logic values of that address line will be copied onto the shift register, pass along to the end with each clock pulse and eventually start to fall out the far end (see Figure 18.7).

Figure 18.7

Data in ⟶

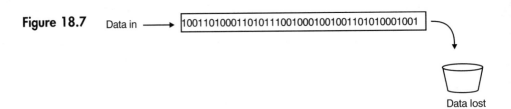

100110100011010111001000100100110101010001001

Data lost

Now if we had four such registers, we could collect data from any four parts of a circuit at the same time. For example, we could monitor the lowest four address lines, which would be called A0, A1, A2 and A3.

In the centre of the register is a window. This means that we can access the centre of the shift register at this point to read off the data and to make comparisons. In Figure 18.8 only the four registers are shown for clarity. A simple arrangement like this would be referred to as a 4 × 16 (four by sixteen) logic analyser. In logic analyser specifications, the number of registers would be spoken of as the number of channels. In real life, we would never find such a simple arrangement of registers. Logic analysers could contain, say, 80 channels, each containing 4096-bit shift registers. This would be referred to as a 80 × 4 kbyte logic analyser. With one this size, we could monitor any 80 different points on a microprocessor-based system. Which points we choose are

Figure 18.8

Four shift registers can make a simple logic analyser

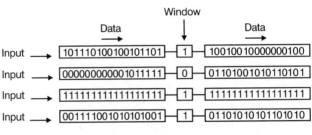

Window

Data ⟶   Data ⟶

Input ⟶ 101110100100101101 ─ 1 ─ 10010010000000100

Input ⟶ 000000000001011111 ─ 0 ─ 01101001010110101

Input ⟶ 111111111111111111 ─ 1 ─ 11111111111111111

Input ⟶ 001110010101010011 ─ 1 ─ 01101010101101010

Window value = 1011

up to us, we could choose the whole of the address bus and the data bus and some control signals or any other points of interest. The choice is entirely ours.

If, in Figure 18.8, the four registers were being used to monitor four address lines, we may be suspicious of the line showing a constant value of logic 1. This may indicate that this line has become short-circuited to a positive power supply, or be disconnected and is floating high. Don't leap off your chair in excitement though – this is only one explanation. It could happen for these reasons or it could be running a part of the program where this would be the expected result.

So what about the window?

In the window, the logic analyser will 'see' a bit of data from each of the channels. We can load the combination that we are searching for. For convenience, we enter the values in hex numbers and as the clock pulses arrive from the microprocessor, the data moves across and is continuously compared with the number we have entered. When a match is found, the clock is switched off and the data is 'captured'. We can now move backwards and forwards along the registers and see the operation of the microprocessor 'frozen' in time.

The benefit of positioning the window in the centre of the shift register is that it allows us to observe the program action before, as well as after, the chosen moment.

Extra facilities

A 'real' logic analyser has some extra facilities, like performing the capture not on the first time our input is seen but after perhaps the 200th occasion to take care of repetitive loops in the program. They also allow a 'don't care' condition on the inputs so in the window of a 20-channel logic analyser we could enter the hex code 7XXX2. This would perform a capture on any data that starts with 7H (0111_2) and ends with 2H (0010_2).

A glitch is a very short duration pulse that can occur in logic circuits, either from external interference or as a result of poor design. They can cause unwanted switching in the logic circuit and cause the microprocessor program to crash. They are exceedingly short, just a few nanoseconds and this makes spotting them very difficult. They are usually too fast for an oscilloscope but some logic probes have a 'glitch-catcher' built in, but they can only tell us that a glitch has occurred, not when it occurred. This is the information that will be needed if we are to track down a design problem.

A logic analyser may miss it because the incoming data is sampled once per pulse and if it misses the glitch it will not be recorded. To overcome this, the logic analyser can use its own internal clock that is

Figure 18.9

Glitch catching

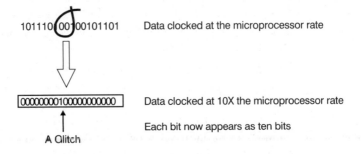

101110 00 00101101   Data clocked at the microprocessor rate

0000000010000000000   Data clocked at 10X the microprocessor rate

Each bit now appears as ten bits

A Glitch

running much faster than the system clock so a single logic one may extend for 10 or 20 bits in the register and a glitch may well be recorded. Figure 18.9 shows the internal clock running at 10 times the microprocessor clock. Some logic analysers have a built-in glitch catcher and use it to capture the correct section of data. As we can see, the logic analyser is a very useful and sophisticated piece of kit. Using it, however, is a slow process. There are lots of connections to be made to the circuit and much sitting and thinking.

Quiz time 18

In each case, choose the best option.

---

1   Damage due to static electricity:

(a) can only occur in the winter.
(b) is best prevented by wearing wet clothing.
(c) is only possible in carpeted areas.
(d) can be reduced by wearing a grounded wrist strap.

---

2   If the two power supplies were connected as in Figure 18.10, the result would be:

(a) smoke pouring from both power supplies.
(b) an output of +5 V but twice as much current.
(c) an output of +10 V.
(d) no output but no smoke either.

Figure 18.10

(a)

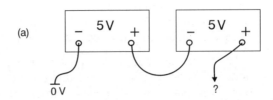

5 V   5 V

0 V   ?

3   The arrow in Figure 18.11 indicates pin:

(a) 2.
(b) 8.
(c) 11.
(d) 16.

Figure 18.11

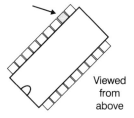

Viewed
from
above

4   A logic probe:

(a) indicates whether your fault-finding technique is based on sound reasoning.
(b) can detect the difference between a disconnection and a grounded connection.
(c) can store a stream of data.
(d) detects the presence of static electricity.

5   A logic analyser quoted as 20 channels x 1024 bits:

(a) will show a four digit hex number in its window.
(b) can monitor any 1024 points at the same time.
(c) would store a total of 1044 bits of data.
(d) can monitor any 20 points at the same time.

Appendix A: Special function register file

Addr	Name	Bit 7	Bit 6	Bit 5	Bit 4	Bit 3	Bit 2	Bit 1	Bit 0	Power-on Reset
BANK 0										
00H		Uses contents of FSR to address data memory (not a physical register)								
01H	TMR0	8-bit real-time clock/counter								xxxxxxxx
02H	PCL	Low order 8 bits of the program counter (PC)								00000000
03H	Status	IRP	RP1	RP0	TO	PD	Z	DC	C	00011xxx
04H	FSR	Indirect Data Memory Address Pointer								xxxxxxxx
05H	PortA	–	–	–	RA4/T-CKI	RA3	RA2	RA1	RA0	---x xxxx
06H	PortB	RB7	RB6	RB5	RB4	RB3	RB2	RB1	RB0/INT	xxxxxxxx
07	–	Unused								
BANK 1										
81H	Option	RBPU	INTEDG	T0CS	T0SE	PSA	PS2	PS1	PS0	11111111
85H	TRISA	–	–	–	PortA data Direction					---11111
86H	TRISB	PortB data Direction								11111111
87H	–	Unused								
88H	EECON 1	–	–	–	EEIF	WRERR	WREN	WR	RD	---0x000
89H	EECON 2	EEPROM control reg. (not a physical register)								--------

The full details (85 pages) of absolutely everything about the PIC16F84A (or any other PIC) can be downloaded from www.microchip.com.

Appendix B: PIC 16CXXX instruction set

Syntax	Description	Status affected
ADDLW k	The contents of the W register are added to an 8-bit number and the result put in the W reg.	C.DC,Z
ADDWF f,d	Add the contents of the W and f registers. If d=0 the result goes to W. If d=1 the result goes to the f register	C,DC,Z
ANDLW k	The contents of the W register are ANDed with an 8-bit number and the result put in the W reg.	Z
ANDWF f,d	AND w with reg f. If d=0 the result goes to W. If d=1 the result goes to the f register	Z
BCF f,b	Bit b in reg f is cleared	
BSF f,b	Bit b in reg f is set	
BTFSS f,b	If bit b in reg f=0, the next instruction in executed. If it is 1 the next instr. is replaced with a NOP	1 or 2 cycles
BTFSC f,b	If bit b in reg f=1, the next instruction in executed. If it is 0 the next instr. is replaced with a NOP	1 or 2 cycles
CALL k	Call subroutine. Return address (PC+1) is pushed onto stack. The 11-bit immediate address is loaded into PC bits <1:0> The upper bits of the PC are loaded from PCLATH <4:3>.	2 cycle
CLRF f	Register f is cleared and Z flag is set.	Z
CLRW	Register W is cleared and Z flag is set.	Z
CLRWDT	Resets watchdog timer and watchdog prescalar	TO, PD
COMF f,d	Contents of 'f' are complemented (0∏1, 1∏0) If d=0 the result goes to W. If d=1 the result goes to f	Z
DECF f,d	Contents of 'f' reduced by 1. If d=0 the result goes to W. If d = 1 the result goes to f	Z
DECFSZ f,d	Contents of 'f' reduced by 1. if d=0 the result goes to W. If d=1 the result goes to f. If result = 1, the next instruction in executed. If it is 0 the next instr. is replaced with a NOP	1 or 2 cycles
GOTO k	The 11-bit immediate address is loaded into PC bits <10:0> The upper bits of the PC are loaded from PCLATH <4:3>.	
INCF f,d	If d=0 the result goes to W. If d=1, result goes to f.	
INCFSZ f,d	Contents of 'f' incremented. If d=0 the result goes to W. If d=1 the result goes to f. If rfesult = 1, the next instruction is executed. If it is 0 the next instr. is replaced with a NOP	1 or 2 cycles

Syntax	Description	Status affected
IORLW k	The contents of the W register are ANDed with an 8-bit number and the result put in the W reg.	Z
IORFWF f,d	The contents of the W register are Inclusive ORed with reg. F. If d=1 resujlt goes back into f	
MOVF f,d	If d=0, contents of f goes to W reg. If d=1 it goes to f	Z
MOVLW k	The 8-bit number k goes into W.	
MOVWF f	Moves data from W register to f register	
NOP	Does nothing – just a time waster (one cycle period)	
RETFIE	Return from interrupt. Top of stack⎾⎿PC, 1⎾⎿GIE	2 cycle
RETLW k	W reg loaded with number , return address⎾⎿PC	2 cycle
RETURN	Return from subroutine. Return address⎾⎿PC	2 cycle
RLF f,d	Contents of 'f' are rotated left one bit via the carry flag. If d=0 the result goes to W. Id=1, result goes back to f. See fig. below	C
SLEEP	Powerdown status bit PD is cleared, Timeout status bit TO is set WDT and prescaler are cleared, oscillator stops and controller goes to sleep.	TO, PD
SUBLW k	W register subtracted from the number k, result goes into W reg. (2's complement method)	C, DC, Z
SUBWF f,d	W register subtracted from the register f. If d=0 the result goes to W. If d=1 the result goes to f. (2's complement method)	C, DC, Z
SWAPF f,d	Upper and lower nibbles of f are exchanged. If d=0 the result goes to W. If d=1 the result goes to f.	
XORLW k	W register contents XOR'ed with the number k, result goes into W reg.	Z
XORWF f,d	W register contents XOR'ed with the register f, if d=0 the result goes to W. If d=1 the result goes to f	Z

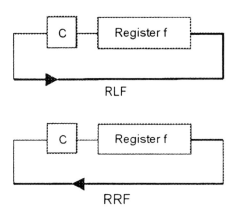

RLF

RRF

Further reading

Bates, M. (2000) The Pic 16F84 Microcontroller. Arnold, London.

Bedford, M. (1996) Jubilee chips, *Computer Shopper*, December.

Bull, M. (1992) *Students' Guide to Programming Languages*. Butterworth-Heinemann, Oxford.

Carthy, J. (1996) *An Introduction to Assembly Language Programming and Computer Architecture*. International Thomson Computer Press.

Crisp, J. (1996) *Introduction to Fiber Optics*. Butterworth-Heinemann, Oxford.

Diefendorff, K., Oehler, R. and Hochsprung, R. (1994) Evolution of PowerPC architecture, *IEEE Micro*, April.

Digital Equipment Corporation (1996) Hardware reference manual of the Digital Semiconductor 21164 Alpha Microprocessor.

Horowitz, P. and Hill, W. (1989) *The Art of Electronics*. Cambridge University Press, Cambridge.

Intel (1988) *Microprocessor and Peripheral Handbook*, Vol. 1.

Krause, J.K. (1997) A Chip off the old block, *BYTE*, November.

Messmer, H.-P. (1995) *The Indispensable Pentium Book*. Addison-Wesley, New York.

Peleg, A. and Weiser, U. (1996) MMX technology extensions to the Intel architecture, *IEEE Micro*, August.

Predko, M. (1998) Programming and customizing the PIC microcontroller. McGraw-Hill, New York.

Wideman, G. (1986) *Computer Connection Mysteries Solved*. H.W. Sams, USA.

www.microchip.com

www.atmel.com

www.intel.com

Quiz time answers

Quiz time 1
   1 (c)
   2 (a)
   3 (c)
   4 (a)
   5 (d)

Quiz time 2
   1 (b)
   2 (c)
   3 (a)
   4 (b)
   5 (d)

Quiz time 3
   1 (d)
   2 (a)
   3 (c)
   4 (b)
   5 (c)

Quiz time 4
   1 (c)
   2 (c)
   3 (a)
   4 (b)
   5 (c)

Quiz time 5
   1 (c)
   2 (a)
   3 (d)
   4 (d)
   5 (c)

Quiz time 6
   1 (d)
   2 (b)
   3 (c)
   4 (b)
   5 (c)

Quiz time 7
   1 (c)
   2 (a)
   3 (d)
   4 (d)
   5 (a)

Quiz time 8
   1 (b)
   2 (a)
   3 (b)
   4 (a)
   5 (c)

Quiz time 9

1 (a)
2 (c)
3 (c)
4 (d)
5 (b)

Quiz time 10

1 (b)
2 (b)
3 (d)
4 (c)
5 (c)

Quiz time 11

1 (a)
2 (c)
3 (b)
4 (a)
5 (d)

Quiz time 12

1 (d)
2 (d)
3 (c)
4 (b)
5 (a)

Quiz time 13

1 (a)
2 (d)
3 (c)
4 (b)
5 (a)

Quiz time 14

1 (c)
2 (a)
3 (b)
4 (a)
5 (c)

Quiz time 15

1 (a)
2 (b)
3 (d)
4 (d)
5 (c)

Quiz time 16

1 (d)
2 (b)
3 (c)
4 (c)
5 (d)

Quiz time 17

1 (b)
2 (a)
3 (d)
4 (d)
5 (c)

Quiz time 18

1 (d)
2 (c)
3 (c)
4 (b)
5 (d)

Index

4004, 151, 159, 160
6500, 164
6502, 162, 163
6800, 160
68000, 166, 167
6801, 164
8008, 160
80386, 147
8048, 164
8051
8080
8080A, 160, 162
8085A, 161
8088, 160

Accumulator, 103
ADC, 240–243
Address buffers, 112
Address decoder, 92–98
Addressing memory, 73, 74, 80, 81, 90–98
Analog to digital conversion, 240–243
APL, 146, 147
Arithmetic and logic unit, 101
ASCII, 247
Assembler, 127

Assembly language, 126, 131
AT90S/LS2343 microcontroller, 206–210

Base of a number system, 16
Basic, 136–139
BCD, 47–48
Benchmarks, 153, 154
Binary coded decimal (BCD), 47–48
Binary, 14–47
    addition, 39, 40
    complementary, 39–43
    converting to other bases, 17–21, 30–36
    subtraction, 40–43
    system, 16, 17
Bistable see Flip-flop
Branch prediction, 176, 180, 187, 190, 195
Burst mode, 174
Bus interface unit, 186
Bus State Controller, 114
Buses, 88
Byte, 22

C, 142–144
C++, 142–144

Cache, 156
Cell, 72, 73
Central Processing unit (CPU), 100, 101
Centronics data transmission, 245, 246
CISC microprocessors, 158, 159
Clock generator, 112
Clock signals, 63, 64, 85–88
Cobol, 139, 140
Compilers, 134
Complementary numbers, 39–43
CPU, 5
Crystal oscillator, 227

DAC, 243, 244
Data buffer, 110
Data transmission, 240, 244
Denary, 15
    converting to other bases, 17–21
Digital to analog conversion, 243, 244
Direct memory Access, 115
Double precision, 47
Doublewords, 178
DRAM see Dynamic RAM
Dynamic RAM, 75, 79, 80

EEPROM see Electrically erasable
        programmable ROM
Electrically erasable programmable ROM,
        77, 78
Emotion engine, 169
Enable input, 59
Endians, big and little, 188, 189
EPROM see Erasable programmable ROM
Erasable programmable ROM, 77–79
Exceptions, 177
Excess 127 notation, 46
Exponent, 45

Faultfinding, 254–257
Fetch-execute, 118, 119
Flag register, 103–106
Flash converter, 240, 241
Flash memory, 81
Flip-flops, 62–63
Floating point numbers, 44–46
Floating point unit, 176, 180
FLOPS, 153
Fortran, 132–136
Full decoding, 95
Future trends, 148, 149

Gamecube, 168, 169
Games machines, 168–172
Gates, 49–60
    AND, 52–53
    ENOR, 58
    EOR, 57–58
    NAND, 54
    NOR, 55, 56
    NOT, 50–52
    OR, 54, 55
    XNOR, 58
    XOR, 57–58
General purpose registers, 106
GFLOPS see FLOPS
Ghost address, 96–98
Gigabyte, 23
GIGO, 119
Glitch, 257, 262, 263

Hexadecimal, 25–35
    converting to other bases, 28–35
    system, 25–27
High level languages, 132–148
Hyper pipeline, 180, 181

Image address, 96–98
Index registers, 109
Input/output devices, 89
Instruction decoder, 101
Instruction register, 101
Instruction set, 122
Instruction unit, 186
Integer unit, 187, 188
Integrated circuit, 5
Interpreter, 137
Interrupts, 112–114, 234

Java, 144

Kilobyte, 22

Labels, 128–129
Languages:
    APL, 146, 147
    Assembly, 126, 131
    Basic, 136–139

C, 142–144
C++, 142–144
Cobol, 139, 140
Fortran, 132–136
Java, 144
Lisp, 145, 146
Machine code, 123–126
Pascal, 141, 142
Prolog, 147, 148
Smalltalk, 139
Large scale integration, 151
Libraries, 134, 135
Linkers, 134, 135
Lisp, 145, 146
Loaders, 134, 135
Logic analyser, 260, 263
Logic gates see Gates
Logic probe, 257, 258
Long word, 22
LSI, 151

Machine code, 123–126
Mantissa, 45
Maths co-processor, 148
Medium scale integration, 151
Megabyte, 23
Memories, 72–83
Memory maps, 81–83
Memory refresh, 115
Micro, 5
Microcomputer, 5
Microcontroller, 5
Micro-OP, 181
Microprocessor system, 3
Microprocessor-based system, 5, 85–98
Microprocessors:
    4004, 151, 159, 160
    6500, 164
    6502, 162, 163
    6800, 160, 162, 163
    68000, 166
    68000, 166, 167
    6801, 164
    8008, 160
    8048, 164
    8080, 99
    8080A, 160, 162
    8085A, 161
    MMX Pentium, 178
    Pentium 4, 179–182

Pentium, 159, 173–182
Power PC, 159, 184–197
Power PC750Cxe, 168, 169
Z8, 164
Z80, 99
Z80180, 99–119, 162
Microprogram, 101, 159
Microsoft Xbox, 1270–172
MIPS, 152, 153
MMX Pentium, 178, 179
Modem, 250
Monitor program, 138, 139
MPC601, 185–192
MPU, 5
MSI, 151

Nintendo Gamecube, 168, 169
Noise, 8-14
    effect of, 14
    partition, 13
    thermal, 13
Non-destructive readout, 126–127
Normalizing, 44

Object code, 127
Object-oriented programming, 142, 143
Octal, 34, 35
Op code, 123
Operand, 123
Optic fibre link, 251, 252
Oscillators, 217

Parity, 236
Partial decoding, 95–98
Pascal, 141, 142
Pentium 4, 179–182
Pentium, 159, 173–182
Performance of microprocessors, increasing,
    154–159
Performance tests:
    FLOPS, 153
    benchmarks, 153, 154
    GFLOPS see FLOPS
    I/O operations, 154
    SPECmark, 154
    TPS, 154
PIC16F84A Option register, 214–216
PIC16F84A pinout, 220

PIC16F84A register file map, 212
PIC16F84A special function register, 213–216
PIC16F84A, 210–217
Pipelining, 156, 157
Playstation 2, 169
Polling interrupts, 235
Power PC 601, 185–192
Power PC, 159, 184–192, 197
PowerPC970, 189–192
Prefetch buffer, 175
Program counter, 107
Programmable ROM, 77–79
Programming a PICF84A, 222–231
Prolog, 147, 148
PROM see Programmable ROM

Quadwords, 178
Quantization error, 242, 243

Radix, 45
RAM cards, 81
RAM, 72–75
Ramp generators, 241, 242
Rapid Execution Engine, 182
Read/write, 65
Redundancy, 236
Register, 64–71
   rotate, 70, 71
   shift, 67–70
Remarks, 130
Reserved words, 129
RISC microcontroller, 206–210
RISC microprocessors, 158, 159
ROM, 76–80
Rotate registers, 70, 71
RS232C and RS423A, 248, 249

Saturation arithmetic, 178, 179
Serial inputs and outputs, 115
Shift registers, 67–71
Signed magnitude numbers, 39
Single precision, 47

Sleep mode, 209, 210, 211
SLSI, 151
Small scale integration, 151
Sony Playstation 2, 169
Source code, 127
SPECmark, 154
SRAM see Static RAM
SSI, 151
Stack pointer, 107, 108
Stack, 108
Startup, 116, 117
Startup address, 82
Static electricity – precautions, 255
Static RAM, 74, 75, 79, 80
Status register, 103–106
Super scale integration, 151
Syntax, 127
System, 1

Terabyte, 23
Test equipment, 257–263
TPS, 154
Tri-state buffer, 59, 60, 65–66
Truth table, 50

UART, 165, 247, 248
ULSI, 151
Ultra scale integration, 151
USART see UART

Vectored interrupts, 235
Very scale integration, 151
VLSI, 151

Wait states, 115
Watchdog timer, 206
Word, 22

Z8, 164
Z80 see Microprocessors
Z80180 see Microprocessors